PRINCIPLES OF HOLOGRAPHY

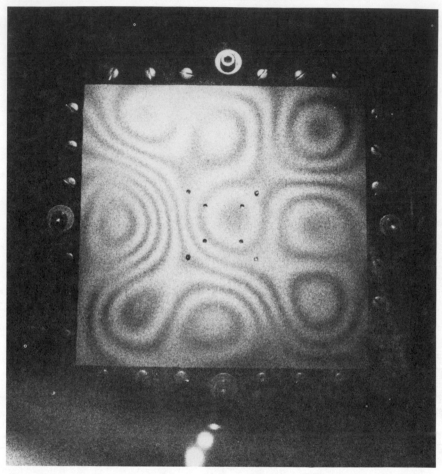

Real-time, stroboscopic, holographic interferometry: An aluminum plate vibrating at a resonant frequency of 2732 Hz. The complex resonant vibration pattern (mode) is captured by holographic interferometry.

Principles of
HOLOGRAPHY

SECOND EDITION

HOWARD M. SMITH

Research Associate
Research Laboratories
Eastman Kodak Company
Rochester, New York

A WILEY-INTERSCIENCE PUBLICATION

JOHN WILEY & SONS

NEW YORK · LONDON · SYDNEY · TORONTO

Copyright © 1969, 1975 by John Wiley & Sons, Inc.

All rights reserved. Published simultaneously in Canada.

No part of this book may be reproduced by any means, nor transmitted, nor translated into a machine language without the written permission of the publisher.

Library of Congress Cataloging in Publication Data:

Smith, Howard Michael, 1938-
  Principles of holography.

  Includes bibliographies.
  1, Holography. I. Title.

QC449.S6 1975 774 75–5631
ISBN 0-471-80341-3

Printed in the United States of America

10 9 8 7 6 5 4 3 2 1

For my daughter Melanie,
   Who will soon be writing books of her own

Preface

In the five years since the publication of the first edition, the science and engineering of holography has done a good deal of "settling in." Needs and aims have been redirected, priorities have been more sharply focused, the range of useful applications has been narrowed, research activity has been narrowed, and, finally, a better understanding of all aspects of holography has been attained. It is this last that is the main reason for this second edition.

Accordingly, the first four chapters, dealing with the history and the basics of holography, are essentially identical with Chapters 1–4 of the first edition. Next comes a wholly new chapter on phase holograms, about which very much has been learned since the first edition was written. Chapter 6 on color holography is not much changed from the old, but Chapter 7 on diffraction efficiency is brand new. Here we present the theory of diffraction efficiency of plane holograms in a much more elegant style than in the first edition. For thick holograms the very elegant and accurate coupled-wave theory is presented in a brief but hopefully continuous and understandable form.

Chapter 8 is also brand new, treating all aspects and sources of noise in holographic imagery. A great deal has been learned in this area since the first edition, both theoretically and experimentally.

Chapter 9 on resolution and Chapter 10 on light sources are essentially the same as their earlier counterparts, but Chapter 11 on materials is entirely new. Since much of the work done in holography since the writing of the first edition has been in the area of materials, this chapter represents an important addition.

The final chapter, 12, is a somewhat enlarged version of Chapter 8 of the first edition. A new section has been added on interferometric fringe interpretation, and the section on holographic microscopy has been expanded. A more elegant derivation of the signal-to-noise ratio of a holographic image storage system is also included.

The style and depth of coverage in this edition should be substantially the same as in the first. The emphasis, however, has changed somewhat in order to give more stress to the practical aspects of holography. Many chemical formulas are given and processing procedures are outlined, for example, and even some recipes for making some recording materials. It is hoped that this stress will lend an overall less theoretical and more practical aspect to this edition.

Howard M. Smith

Rochester, New York
February, 1975

Preface to the First Edition

The field of holography has advanced rapidly since its renaissance in the period between 1962 and 1964. More than 500 papers and articles have been written on the subject since its inception in 1948—the large majority published since 1964.* This rate of publication has led to duplication, incomplete treatments, erroneous statements, and a careless nomenclature. Within this great volume of published material, however, are several major papers that treat one aspect or another of the science of holography in a definitive manner. Because of this somewhat confusing proliferation of writings on the subject, one of the main purposes of this book is to present a unified and complete treatment under one cover. To do this I have presented with little or no change some of the more important approaches to the various subjects. Of course, the whole idea of off-axis holography must be credited to Emmett Leith and Juris Upatnieks of the University of Michigan.† The work in Section 5.3, which deals with the effects of the recording medium on image resolution, has been borrowed almost entirely from the two papers by Raoul Van Ligten of the American Optical Company.‡ Section 5.4 deals with the third-order aberrations of the holographic process and has been taken virtually unchanged from the paper by R. W. Meier of Xerox. § Finally, Section 6.1.2 is made up entirely of the work presented in the papers by Kaspar and Lamberts,‖ and Kaspar, Lamberts, and Edgett¶ of Eastman Kodak. The rest of the material has been gleaned from other publications or has been worked out by me, usually in a very simple fashion.

Some of the terminology and usage associated with this field has become inaccurate and inconsistent, due in part, I suppose, to the large volume of

* J. N. Latta, *J. Soc. Motion Picture Television Engrs.*, **77,** 422 (1968).
† E. N. Leith and J. Upatnieks, *J. Opt. Soc. Am.*, **52,** 1123 (1962).
‡ R. F. Van Ligten, *J. Opt. Soc. Am.*, **56,** 1, 1009 (1966).
§ R. W. Meier, *J. Opt. Soc. Am.*, **56,** 219 (1966).
‖ F. G. Kaspar and R. L. Lamberts, *J. Opt. Soc. Am.*, **58,** 970 (1968).
¶ F. G. Kaspar, R. L. Lamberts, and C. D. Edgett, *J. Opt. Soc. Am.*, **58,** 1289 (1968).

papers published each month in the technical journals. I have tried to use a "correct" terminology and I hope that it will be generally accepted. I have never employed the phrase "to reconstruct an image," since in holography it is the wavefront that is being reconstructed and not the image. Also, I have never employed the phrases "to reconstruct the object" or, worse yet, "to reconstruct the hologram." One illuminates the hologram in such a way that a wavefront is reconstructed which can be used to form an image of the object. This is the usage I have tried to maintain. The two images associated with the holographic process are almost universally referred to as the real and virtual images, even though this designation does not specify completely the image to which we refer. The "real" image, for example, may very well be virtual, which can only lead to confusion. I have therefore consistently designated the two holographic images as the *primary* and *conjugate* images.

These are small points, to be sure, but I feel that a careless and loose terminology tends to degrade the field and confuse those not familiar with conventional usage.

I have tried to keep the physics as simple as possible, so that the reader can get a clear insight into what is going on in any given situation. The field of holography is not a complex or intricate one in any case—most things can be understood from a fundamental viewpoint and I have attempted to stress this in the treatment of each topic. The book is written so that any photographic or optical engineer will be able to learn the whys and wherefores of any aspect of holography and thus be able to deal intelligently with any questions that might arise, regardless of whether he is working in the laboratory or making management-level decisions concerning holography. It is assumed that the reader has a knowledge of several aspects of optics, such as modulation transfer functions, spatial frequency, phase, coherence, diffraction, and so on. Some of these concepts are treated only briefly when they are introduced.

Finally, I should warn the reader that some material is repeated from time to time for the sake of completeness. The lensless Fourier transform hologram is described in Chapter 3 and again in Chapter 6. The equation describing the exposure at the hologram phase has been included almost every time a new topic is introduced. The integral specifying the object field at the hologram plane appears several times throughout the text. These things, and possibly some others, have been repeated so that each chapter, and even each topic, can be read and understood almost independently of the others.

Howard M. Smith

Rochester, New York
June, 1968

Contents

1. Historical Introduction   **1**

2. Basic Arrangements for Holography   **13**

    2.0   Introduction   13
    2.1   Basic Description of Holography   13
    2.2   Arrangements for Recording Plane Holograms   24
    2.3   Arrangements for Recording Volume Holograms   27

3. General Theory of Plane Holograms   **33**

    3.0   Introduction   33
    3.1   Notation   33
    3.2   Analysis   34
           3.2.1   Fresnel Holograms   34
           3.2.2   Fraunhofer Holograms   42

4. General Theory of Volume Holograms   **48**

    4.0   Introduction   48
    4.1   Fringes in Three Dimensions   49
    4.2   Diffraction from a Three-Dimensional Grating   52
           4.2.1   The Three-Dimensional Grating Equation 52
           4.2.2   The Pupil Function $G(x, z)$ 55
           4.2.3   The Bragg Condition   57
           4.2.4   Wavelength Change   62

4.2.5   Orientation Sensitivity   63
4.2.6   Wavelength Sensitivity   68
4.3   Reflection Holograms   71

5. Phase Holograms   76

5.0   Introduction   76
5.1   Analysis   77
5.2   Diffraction Efficiency and Noise   80
5,3   Methods for Producing Phase Holograms   82
5.3.1   Direct Bleaching   82
5.3.2   Dichromated Gelatin   83
5.3.3   Thermoplastics   84
5.3.4   Reversal Bleaching   86
5.4   Conclusions   89

6. Color Holography   91

6.0   Introduction   91
6.1   Analysis   94
6.2   Plane Hologram Techniques   97
6.2.1   Three-Reference-Beam Method   97
6.2.2   Spatial Multiplexing   97
6.2.3   Coded Reference Beam   98
6.3   Volume Hologram Techniques   100
6.3.1   Transmission Holograms   100
6.3.2   Reflection Holograms   101
6.4   Diffraction Efficiency   102
6.5   Colorimetry   104

7. Diffraction Efficiency   107

7.0   Introduction   107
7.1   Plane Holograms   109
7.1.1   Amplitude Holograms   109
7.1.2   Phase Holograms   112
7.2   Volume Holograms   113
7.2.1   Coupled-Wave Theory   113
7.2.2   Transmission Holograms   115
7.2.2.1   Pure Phase Holograms   115
7.2.2.2   Pure Phase Holograms with Loss   116

7.2.2.3   Amplitude Holograms   117
7.2.2.4   Mixed Holograms   119
7.2.3   Reflection Holograms   119
7.2.3.1   Pure Phase Holograms   119
7.2.3.2   Amplitude Holograms   120
7.2.4   Summary   121
7.3   Equivalence of Thick and Thin Holograms   122

8. Noise   125

8.0   Introduction   125
8.1   Granularity   126
8.2   Scattering from the Support Material   132
8.3   Phase Noise   132
8.4   Nonlinearity Noise   135
8.5   Speckle Noise   136

9. Holographic Image Resolution   143

9.0   Introduction   143
9.1   Reference and Illuminating Beam Source Size   143
9.2   Reference and Illuminating Beam Bandwidth   152
9.3   Effect of the Recording Medium   155
9.3.1   Introduction   155
9.3.2   Fourier Transform Holograms   156
9.3.3   Fresnel Holograms   159
9.4   Third-Order Aberrations   169
9.4.1   Introduction   169
9.4.2   Analysis   169
9.4.2.1   Magnification   170
9.4.2.2   Third-Order Aberrations   172

10. Light Sources for Holography   178

10.1   The Gas Laser   178
10.1.1   Temporal Coherence   178
10.1.2   Spatial Coherence   185
10.2   The Solid State Laser   187
10.3   The Line Source   190

11. Recording Materials for Holography   **196**

    11.0 Introduction   196
    11.1 General Requirements   196
    11.2 Hologram Types   199
    11.3 Silver Halide Photographic Emulsions   202
    11.4 Photochromics   205
    11.5 Photoresists   207
    11.6 Photopolymers   208
    11.7 Dichromated Gelatin   210
    11.8 Thermoplastics   213
    11.9 Electro-Optic Crystals   216
    11.10 Other Holographic Recording Materials   217

12. Applications for Holography   **220**

    12.0 Introduction   220
    12.1 Holographic Interferometry   221
          12.1.1 Introduction   221
          12.1.2 Single-Exposure Holographic Interferometry   222
          12.1.3 Double-Exposure Holographic Interferometry   227
          12.1.4 Fringe Interpretation   232
          12.1.5 Time-Average Holographic Interferometry   234
          12.1.6 Contour Generation   236
    12.2 Correcting Aberrated Wavefronts with Holograms   238
    12.3 Matched Filtering and Character Recognition   242
    12.4 Holographic Microscopy   247
    12.5 Imagery Through Diffusing Media   254
    12.6 Three-Dimensional Observation   256
    12.7 Information Storage   257
          12.7.1 Introduction   257
          12.7.2 Coherent and Incoherent Superposition   257
          12.7.3 Storage Capacity   260
    12.8 Ultrasonic Holography   264

Appendix Fourier Transforms with Lenses   **268**

    A.0 Introduction   268
    A.1 Analysis   268

Index   **275**

PRINCIPLES OF HOLOGRAPHY

1 Historical Introduction

Ever since 1900, man has been able to record and retain as a permanent record almost any scene that his eyes perceived, through the process of photography. The optical lens had been invented and used several centuries prior to this date, and the formation of optical images with lenses was well understood by 1900. With the invention of the photographic process, the importance of the lens in scientific investigation was greatly enhanced. The fortunate combination of lens and photographic emulsion made possible the charting of stars, planets, and galaxies; the recording of optical spectra; the picturing of minute microscopic specimens; the storage of large amounts of data in the form of small recorded images; and myriad other uses. Because of the vast scope of its scientific importance, the science of photography has advanced steadily over the past seventy or more years; even today new and important uses are being found.

Now science has at its disposal a new method of forming optical images: holography.

Holography is a relatively new process that is similar to photography in some respects but is nonetheless fundamentally different. Because of this fundamental difference, holography and photography will not be competing in the same areas. There are several applications for which holography is more suitable than photography, while most of the more important uses for photography remain unchallenged. Furthermore, there are several tasks that can be performed with holography but not at all with conventional photography.

In order to point out the fundamental differences between holography and photography, we should understand in a general way how each works.

Photography basically provides a method of recording the two-dimensional irradiance distribution of an image. Generally speaking, each "scene" consists

1

of a large number of reflecting or radiating points of light. The waves from each of these elementary points all contribute to a complete wave, which we will call the "object" wave. This complex wave is transformed by the optical lens in such a way that it collapses into an image of the radiating object. It is this image which is recorded on the photographic emulsion.

Holography, on the other hand, is quite different. With holography, one records not the optically formed image of the object, but the *object wave itself*. This wave is recorded in such a way that a subsequent illumination of this record serves to *reconstruct* the original object wave, even in the absence of the original object. A visual observation of this reconstructed wavefront then yields a view of the object or scene that is practically *indiscernible from the original*. It is thus the recording of the object wave itself, rather than an image of the object, which constitutes the basic difference between conventional photography and holography.

A brief description of how the object wave is recorded will be useful before tracing the history of holography. One starts with a single monochromatic beam of light that has originated from a very small source. The requirements that this beam of light be monochromatic and that it originate from a small source together form the condition that the light be *coherent*. The requirement of coherence means that the light should be capable of displaying interference effects that are stable in time. This single beam of light is then split into two components, one of which is directed toward the object or scene; the other is directed to a suitable recording medium, usually a photographic emulsion. The component beam that is directed to the object is scattered, or diffracted, by that object. This scattered wave constitutes the object wave, which is now allowed to fall on the recording medium. The wave that proceeds directly to the recording medium is termed the *reference wave*. Since the object and reference waves are mutually coherent, they will form a stable interference pattern when they meet at the recording medium. This interference pattern is a complex system of fringes—spatial variations of irradiance that are recorded in detail on the photographic emulsion. The microscopic details of the interference pattern are unique to the object wave—different object waves (objects) will produce different interference patterns. The detailed permanent record of this interference pattern on the photographic emulsion is called the "hologram," from which the word "holography" is derived. This photographic record, or hologram, now consists of a complex distribution of clear and opaque areas corresponding to the recorded interference fringes. When the hologram is illuminated with a beam of light that is similar to the original reference wave used to record the hologram, light will only be transmitted through the clear areas, resulting in a complex transmitted wave. However, because of the action of the recorded interference fringes, this transmitted wave conveniently divides into three separate components, *one of which exactly duplicates the original object wave.* Thus by viewing this reconstructed wavefront, one sees an exact replica of the original object, even though the object is not present during the reconstruction process. Thus holography is a two-step process by which images can be formed. In the first step, a complex interference pattern is recorded and becomes the hologram. In the second step, the hologram is illuminated in such a way that part of the transmitted light is an exact replica of the original object wave. The fundamental difference between holography and conventional photography is now quite evident.

This method of optical imagery is not really new. Nearly three decades ago British research scientist Dennis Gabor first conceived of, as he called it, "a new two-step method of optical imagery" [1]. However, it is only in the past few years that the method has become widely known and used. The modern renaissance in holography had to await the general availability of the laser with the great temporal and spatial coherence of its light, but the really significant contributions to Gabor's original idea were more basic in nature.

The general idea of this two-step imaging process was suggested to Gabor by considerations of Bragg's x-ray microscope [2]. Bragg had been able to form the image of a crystal lattice by means of diffraction from the photographically recorded x-ray diffraction pattern of the lattice. The basic idea behind Bragg's method is a double-diffraction process, which is the crux of the holographic process. Image formation by double diffraction becomes clear if we note that the field diffracted by an object can be represented as a Fourier transform of the light distribution at the object [3]. Thus the second diffraction becomes a Fourier transform of the Fourier transform of the object, which is an image of the object itself. This means that diffraction from the hologram will reproduce the object wave, provided that *all the amplitudes and phases of both diffractions are preserved.*

It was just this question of phase preservation that represented the basic limitation to Bragg's method. Since he was able to record only irradiances, phase information was discarded. He was thus limited to applying the method only to a restricted class of objects, such as crystal lattices, for which the absolute phase of the diffracted field could be predicted. It was the preservation of the phase information, or at least rendering the recording of the phase unimportant, which represents the crux of Gabor's method. Bragg was able to circumvent the phase problem by using a class of objects for which a known phase change occurs between the incident and diffracted radiation. By using crystals having a center of symmetry, all the scattered radiation was either in phase or 180° out of phase with the incident radiation. Hence by recording the diffraction pattern photographically and therefore recording only irradiance, *no* phase information was lost and another diffraction of this photograph restored the original object wave.

Buerger [4] was later able to extend Bragg's method to crystals that did not have a center of symmetry but nevertheless had approximately known phase changes. By inserting appropriately placed "phase shifters," he was able to reconstruct the true wave of the original object.

Gabor [1] was able to extend these ideas by reasoning that the phase of the diffracted wave could be determined by comparison with a standard reference wave. He added to the diffracted wave a strong, uniform radiation, the amplitude of which is modulated by the diffracted radiation, provided the two are coherent with respect to one another. A photograph of the resulting modulated diffracted wave constitutes the hologram. Diffraction of radiation from the hologram then gives the second diffraction, resulting in the reconstruction of the original wavefront.

The situation, however, is not quite that simple, as Gabor recognized from the start. The photographic plate records the modulated amplitude of the diffracted wave, but it still does not record its absolute phase. Gabor made the diffracted wave weak with respect to the reference wave, so that the phase of the resultant wave (diffracted plus reference) was always approximately that of the reference wave. Thus the diffracted wave could be considered to have the phase of the reference wave and vary only in amplitude. This approximation results in the production of two waves—the original wave diffracted by the object and a "twin wave" that has the same amplitude, but opposite phase, relative to the reference wave. Efforts to eliminate the effects of the twin wave constituted a large portion of the early work in wavefront reconstruction, or "holography" [1, 5, 6]. The recent complete elimination of this problem [7] was one of the major reasons for the dramatic and widespread resurgence of interest in holography.

Gabor first demonstrated the feasibility of the holographic method in 1948 [8]. The radiation he used was light. The necessary monochromaticity was obtained by filtering the radiation from a mercury arc, and the required spatial coherence was obtained by illuminating pinholes. His primary objective was to increase the resolution of the electron microscope. The theoretical resolution limit of the electron microscope at that time was about 5 Å, which was determined by a compromise between diffraction and the spherical aberration of electron objectives. Since at that time it seemed unrealistic that any further correction of the aberrations of electron objectives was feasible, Gabor reasoned that by making the hologram with electrons and reconstructing with light, an aberration-free image could be obtained, provided the light optics precisely matched the aberrations of the electron optics. The procedure he suggested was to put the object in the electron beam, just in front of a reduced image of the electron source. A hologram formed by the electrons diffracted from the object was recorded on a photographic plate some distance beyond the object. The hologram was then scaled up optically in the ratio of the light wavelength to the electron wavelength and illuminated with a light wave with the same aberration as the electron wave, scaled also in the same ratio. Theoretically, then, the object was visible through the hologram, in the original position and magnified by this same ratio of light to electron wavelength, or about 100,000×.

Because of certain technical difficulties, such as mechanical and electrical stability, the method did not succeed. Haine and Dyson subsequently suggested an improved arrangement that increased the usable field, thus relaxing the electrical stability requirements [9]. By this method, the "transmission method," lenses were used between the object and the hologram in order to magnify the diffraction pattern. This increased the effective resolution of the photographic plate. The required apparatus was essentially identical with the classical electron microscope, thus facilitating the location of the object.

Efforts by Haine and Mulvey to implement this method were again frustrated by practical limitations [10]. They managed to obtain diffraction resolutions of about 6 Å by eliminating as many of these problems as possible but could not go further because of difficulty in holding the specimen stationary in the electron beam. It was necessary to hold the object stationary relative to the objective lens to within a few angstrom units during exposure, a period of several minutes. Further difficulties were encountered in the optical reconstruction stage.

The most serious problem in reconstruction was the disturbance created by the twin wave. Because of the uncertainty of π in the recording of the phase in the hologram, there are two possible objects giving rise to the same intensity distribution in the hologram. One of these is the original object, the other a virtual object located symmetrically behind the source. Upon reconstruction, waves from both objects are formed. Therefore, in viewing the image of the real object, one has to look through an out-of-focus background image of the virtual object, a most annoying disturbance. Gabor [1] noted that if a condenser system is used to form a reduced image of the source, the twin image will be severely affected by the aberrations of the condenser system. Thus it will appear blurred while the image of the original object will appear sharply defined, the aberrations having been compensated for in this image during reconstruction. Gabor [11] also suggested a method whereby an obscuring mask is placed at a suitable formed image of the point source during reconstruction. Thus most of the wave containing the information of the real object is passed, but most of the background is suppressed. Bragg and Rogers [5] suggested that since the reconstruction is really an in-focus image of the object plus a hologram of the conjugate object, a second hologram can be made of the conjugate object and subtracted from the original. The subtraction is performed by placing the second hologram in contact with the original image. The high-transmission portions of the secondary hologram fall on the high-density regions of the unwanted image and vice-versa, so that the back-ground becomes uniform. The great precision required for registering the two plates prevented complete success, but the effects of the twin image were reduced [5, 6].

Paralleling these early efforts to utilize holography for the improvement of electron microscopy were efforts by several workers to produce x-ray holograms. El Sum [6] produced an artificial x-ray hologram of a thin wire by photographing a published picture of the x-ray diffraction fringes of the wire. He managed to obtain a reasonable reconstruction, using light, from this hologram, at least proving the feasibility of x-ray holography. Baez [12] did a theoretical study of the problems and also concluded that holograms and reconstructions with x-rays are feasible. He does note, however, that because of film resolution and source size limitations, useful resolution might be achieved with visible light. Further work on x-ray holography is still awaiting a small, monochromatic source of x-rays.

Aside from efforts by Rogers [13, 14] and Kirkpatrick and El Sum [15] to provide more satisfying conceptual explanations, the subject of holography lay dormant for almost a decade. Brief explanations of the principle published in a few optical textbooks [16, 17] represented about all of the published work on the subject for this long period of time. The most serious limitations of the method, which led to the dying out of interest, were the lack of an intense, coherent source in either the x-ray or optical region of the spectrum and the disturbing presence of the twin wave. Exposure times of the order of one hour were not unusual [6], and resolution in the reconstruction never reached theoretical predictions.

Interest began reviving in the field when Leith and Upatnieks [7] demonstrated a method for the complete elimination of the twin wave by a fairly simple means. Describing the holographic process from a communication-theory viewpoint, these authors realized that if the signal information (wavefront diffracted from the object) could be put on a carrier frequency (off-axis reference wave), the two reconstructed waves would then represent the sidebands of the process and be physically separated from each other. From an optical viewpoint, this means that if the wave diffracted from the object is made to interfere with a reference wave that is off-axis, rather than in line, the hologram will then be a grating like structure. Reconstruction will yield two waves representing the two first orders of the grating. One of these waves is then the same as the original wave from the object, and the other is the unwanted twin wave. Thus a physical separation in space of the two waves is achieved and the disturbing effects of the twin waves are eliminated. Gabor [1] had noted from the start that this turn of events would probably occur when he said, "It is very likely that in light optics, where beam splitters are available, methods can be found for providing the coherent background which will allow better separation of object planes and more effective elimination of the effects of the 'twin wave' than the simple arrangements which have been investigated." Baez [12] came very close to introducing the off-axis concept in 1952 when he noted that "in an analogous way a diffraction grating forms a virtual image of a source," while explaining how a hologram forms an image by diffraction.

Thus the idea of Leith and Upatnieks revived interest in holography. Many people began taking note and trying a few simple experiments. The "twin image" would no longer be termed the "unwanted image" but would prove to be a useful adjunct of the holographic process. Although the idea of an off-axis reference wave was a definite advance, there were still problems with dust and imperfections in the optical components. The slightest speck of dust on the lenses or mirrors gave rise to its own hologram, reducing the effective aperture of the hologram and reconstructing itself as noise. El Sum [6] and others had taken great pains to remove this source of noise by rotating as many of the components as possible, thus smearing out the holograms of the nonstationary dust particles. This method was successful to some degree, but extremely impractical. Later developments would eliminate this problem also.

Another advantage of the off-axis reference beam method is elimination of critical film processing. The original method required that the hologram be processed as closely as possible to a gamma (contrast) of two for linear transfer of exposure to amplitude transmission. In the Gabor technique, any nonlinearities in the transmission-exposure transfer resulted in decreased image contrast due to the background light level. In the new technique, any nonlinearities of the recording medium result mainly in higher diffraction orders. These higher orders are diffracted at angles larger than the first-order wave; thus nonlinear recording has little effect on the desired image. Curiously enough, processing the film to a high contrast with this new method, thus increasing the nonlinearity of the recording, is actually somewhat beneficial. A high-contrast hologram results in a brighter image, but with little disturbing effect from the higher-order terms.

The off-axis reference beam method also resulted in the elimination of the effects of self-interference between different points of the object. In the earlier method this self-interference resulted in a veiling glare around the image. In the new method, this noise term can be avoided completely.

Finally, the new method made reconstruction possible for objects that did not transmit a large portion of the incident wave and also for continuous-tone objects. In the earlier technique, it was necessary that the major portion of the light passing the object not be diffracted. This undiffracted light is then only slightly modulated by the light diffracted from the object. In this way the loss of phase information in the recording process is rendered negligible; the resultant phase of the total disturbance at the hologram recording plane is practically that of the background wave. When the hologram is illuminated with the background wave alone, the phase of the original total disturbance is approximated and a recognizable reconstruction results. In the new method, the phase of the total disturbance at the hologram is recorded as a phase modulation of a grating like fringe pattern. This permits reconstructions of a wholly new class of objects that do not transmit a large portion of the incident wave, such as transparent letters on an opaque background and continuous-tone objects.

Thus the new method introduced by Leith and Upatnieks eliminated many of the annoying features of the original method, but the real renaissance of holography had to await two other important advances which were not long in coming.

About the same time (1962) that Leith and Upatnieks were introducing the off-axis reference beam method, people were beginning to make and use a radically different light source that would prove to be eminently suitable for holography. Thus the invention of the gas laser coincided nicely with the revival of interest in holography. The laser was capable of producing very intense monochromatic radiation in regions of the spectrum that could be recorded photographically. Because of the highly coherent nature of the light from a laser, it can be focused down to an arbitrarily small spot; hence source size no longer limits the attainable resolution in a holographic image. The monochromaticity of the laser allows for full utilization of the off-axis recording scheme, since now many more interference orders (fringes) can be recorded. This yields much higher resolutions than had previously been obtained. Also, there are no longer any restrictions on the size of the object to be used; holograms can now be made of very large objects.

The advent of the gas laser made possible still another important advance, again introduced by Leith and Upatnieks. In 1964 these authors introduced the concept of diffuse illumination holography [18]. Before this, the only holograms that had been made were of thin transparent objects. Holograms of these kinds of objects often consisted of nearly recognizable shadowgrams of the object. Thus a small region of the hologram would bear almost a one-to-correspondence with a small region of the object. Viewing the images formed with this type of hologram required some additional optical components, since an observer viewing a specularly illuminated transparency will see, for the most part, only that portion of the transparency which lies on a line between the light source and his eye pupil. Hence without optical aids, only a small portion of the image can be viewed at one time. On the other hand, if the transparency is illuminated diffusely, it can be viewed in its entirety with the eye in one location. This, then, is the idea that Leith and Upatnieks introduced into holography in 1964. By placing a diffuser, such as an opal glass, behind the object, a hologram is formed of both the diffuser and object. In this way it is possible to view the image formed from the hologram by merely looking through the hologram as a window. The ability to view the images formed from this type of hologram is not the only advantage. There are several even more interesting.

Because the object is diffusely illuminated, there are no longer any recognizable shadowgrams of the object on the hologram. The light scattered from each point of the object spreads out so as to cover the entire photographic plate. This means that there is no longer a one-to-one correspondence between a portion of the hologram and a region of the object. The information of any single object point is recorded over the whole plate. Because of this, only a small portion of the hologram is required to form an image of the whole object. In fact, in viewing the hologram with the unaided eye, only a portion of it roughly the size of the eye pupil is actually used. If the hologram were to be broken, scratched, torn, or damaged, it would still be possible to form an image of the complete object, though some resolution would be lost.

A further advantage of holograms of diffusely illuminated objects is that dirt or scratches on the mirrors, beamsplitters, and/or lenses used in making the holograms no longer represent the problem they did in the earlier types. Most of these imperfections are, to a great extent, smeared out over the whole plate and thus have a negligibly small effect on the reconstruction.

Perhaps the single most important aspect of this technique is the ability to record holograms of diffusely reflecting, three-dimensional objects. Gabor had recognized from the beginning that a hologram of a three-dimensional object should be capable of forming a three-dimensional image [1]. El Sum [6] and Rogers [14] managed to make holograms of objects in depth, even with the limited coherence lengths of the light sources at their disposal. The gas laser and the diffuse illumination concept made possible the formation of truly striking three-dimensional images. Since the object is diffusely illuminated or diffusely reflecting, light from a large range of perspectives reaches the photographic plate. An observer viewing the image formed by this hologram can move his head and see around foreground objects, just as if he were viewing the original object; he sees a true three-dimensional image. It is necessary for him to refocus his eyes, depending on whether he is viewing a near or far object point.

The new diffuse-illumination holograms also lend themselves very nicely to the superposition of more than one hologram on a single photographic plate. Rogers [14] had made two holograms on a single plate by double exposure, but they were of two objects suitably situated so that the hologram of one did not obliterate too great a portion of the hologram of the other. Suitable positioning of the objects was necessary with the Gabor-type hologram since the information about an object was fairly well localized in the region surrounding the geometrical shadow of the object. Since this is not the case with diffuse-illumination holograms, it is a simple matter to make multiple holograms on a single plate and still obtain high-quality reconstructions. There is still a limitation on object position, however. To prevent the various reconstructions from each hologram from falling on top of one another, it is necessary that the hologram of each object be recorded with a different angle between the object beam and reference beam. In this way the reconstruction of each object wavefront will be traveling in a different direction and hence be separated in space. Each reconstruction can thus be viewed separately.

This concept led Leith and Upatnieks to propose a method of multicolor wavefront reconstruction [18]. By illuminating a colored object with coherent light in each of the three primary colors, each with its own reference beam, three holograms will be recorded. Reconstruction is accomplished by re-illuminating with the three reference beams, and a full-color object wavefront results. The scheme has been demonstrated, but better methods have evolved. They will be discussed in Chapter 7.

At about the same time that Leith and Upatnieks were advancing the field of holography at a great rate, Denisyuk [19] proposed an idea that proved to be of fundamental significance. He suggested that the wavefront from an object traveling in one direction be made to interfere with a coherent reference beam *traveling in the opposite direction* in a three-dimensional recording medium. In this way a standing wave pattern is set up in the recording medium which is uniquely related to the object wavefront. The medium therefore records a series of surfaces separated by one-half the recording wavelength in the medium. These surfaces, under appropriate conditions, are just the antinodal surfaces of the object wavefront. In this way the actual wavefronts of the object beam are recorded. The recording medium is processed so that a change in the dielectric constant occurs where the exposure is high. Reillumination with the reference wave alone yields a reflected wave that is an exact replica of the object wave. Denisyuk's idea was thus a fruitful combination of Lippmann's [20] color photographic process and the hologram method of Gabor [1]. This idea was later investigated in detail by van Heerden [21] and has proved to be an important aspect of modern holography.

The holographic method differs significantly from the conventional photographic process in several basic respects and has distinct advantages in many areas. The most obvious advantage of holography is the ability to store enough information about the object in the hologram to produce a true three-dimensional image, complete with parallax and large depth of focus. There has been a great deal of work done in an attempt to produce three-dimensional images using conventional photographic techniques. These methods have been only partially successful because of the limited depth of field and restricted viewing conditions. An observer viewing a stereo pair, for example, cannot move his head from side to side and look behind foreground objects, as he can with a hologram. The lenticular-type three-dimensional photograph allows limited parallax, but has a rather severe depth limitation. The hologram, on the other hand, has a field of view that is limited in general only by the resolution of the recording medium. The depth of field recorded in a hologram is limited only by source bandwidth. Thus, if a hologram is made of a three-dimensional object, it is equivalent to many conventional photographs, each taken from a different point of view and each focused at a different depth. Subsequent viewing of the hologram image at different depths requires only a refocusing of the viewing system. Hence it is fair to say that one hologram is worth a thousand pictures!

The quality of a holographic image is less sensitive to the characteristics of the recording medium than a photograph. Holograms made on high-contrast material reproduce tonal variations of the object over a wide range. Nonlinear recording has only a small effect on the final image. Also, imperfections in the emulsion, such as scratches, have very little effect on the final image. Indeed, a modern hologram is so redundant that only a small fraction of the holographic record is necessary to form a complete image.

Because of these basic differences between holography and conventional photography, many interesting and novel applications have been proposed. Few of these have been put to commercial use as yet, but the field is still young.

REFERENCES

[1] D. Gabor, *Proc. Roy. Soc.* (London), Ser. A, **197,** 454 (1949).

[2] W. L. Bragg, *Nature*, **149,** 470 (1942).

[3] F. Zernike, *Ned. Tijdschr. Natuurk.*, **9,** 357 (1942).

[4] M. J. Beurger, *J. Appl. Phys.*, **21,** 909 (1950).

[5] W. L. Bragg and G. L. Rogers, *Nature*, **167,** 190 (1951).

[6] H. M. A. El Sum, *Reconstructed Wavefront Microscopy*, Ph.D. Thesis, Stanford Univ., November, 1952.

[7] E. N. Leith and J. Upatnïeks, *J. Opt. Soc. Am.* **52,** 1123 (1962).

[8] D. Gabor, *Nature*, **161,** 777 (1948).

[9] M. E. Haine and J. Dyson, *Nature*, **166,** 315 (1950).

[10] M. E. Haine and T. Mulvey, *J. Opt. Soc. Am.*, **42,** 763 (1952).

[11] D. Gabor, *Proc. Phys. Soc.* (London), **B64,** 449 (1951).

[12] A. V. Baez, *J. Opt. Soc. Am.* **42,** 756 (1952).

[13] G. L. Rogers, *Nature*, **166,** 237 (1950).

[14] G. L. Rogers, *Proc. Roy. Soc.* (Edinburgh), **A63,** 14 (1952).

[15] P. Kirkpatrick and H. M. A. El Sum, *J. Opt. Soc. Am.*, **46,** 825 (1956).

[16] R. S. Longhurst, *Geometrical and Physical Optics*, Longmans, Green and Co., Ltd., London, 1957, p. 301.

[17] M. Born and E. Wolf, *Principles of Optics*, The Macmillan Co., New York, 1959, p. 453.

[18] E. N. Leith and J. Upatnieks, *J. Opt. Soc. Am.*, **54**, 1295 (1964).

[19] Y. N. Denisyuk, *Soviet Phys.—Doklady* **7**, 543 (1962).

[20] G. Lippman, *J. Phys. Radium*, **3**, 97 (1948).

[21] P. J. van Heerden, *Appl. Opt.* **7**, 393 (1963).

2 Basic Arrangements for Holography

2.0 INTRODUCTION

We begin our discussion of holography by describing some of the general arrangements currently used for recording and reconstructing. We restrict our discussion to the off-axis type of hologram first described by Leith and Upatnieks [1]. Using this scheme, the object and reference beams are coincident at the recording medium, arriving from substantially different directions. This is achieved in practice by placing the object laterally some distance away from the source of the reference beam. The object is, of course, illuminated with a beam of light from the same source that provides the reference beam. As described in the Chapter 1, the recording medium records the two-beam interference pattern; the precise details of the pattern are unique to the object used. This record is now called the hologram of the object, and when it is illuminated with a single beam of light similar to the original reference wave, the hologram diffracts the light in such a way as to reconstruct the object wave.

2.1 BASIC DESCRIPTION OF HOLOGRAPHY

Conceptually, the simplest form of an off-axis hologram is one for which the object is just a single, infinitely distant point, so that the object wave at the recording medium is a plane wave (Fig. 2.1a). If the reference wave is also plane and incident on the recording medium at an angle to the object wave, the hologram will consist of a series of Young's interference fringes. These recorded fringes are equally spaced straight lines running perpendicular to the plane of the figure. Since the hologram consists of a series of alternately clear

13

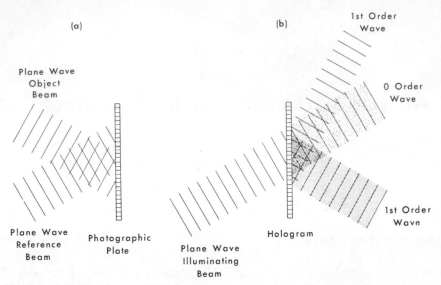

Fig. 2.1 The basic plane wave hologram. (*a*) Recording the two-beam interference pattern. (*b*) The diffraction of a plane wave from the recorded interference pattern.

and opaque strips, it is in the form of a diffraction grating. When the hologram is illuminated with a plane wave (Fig. 2.1*b*), the transmitted light consists of a zero-order wave traveling in the direction of the illuminating wave, plus two first-order waves. The higher diffracted orders are generally missing or very weak since the irradiance distribution of a two-beam interference pattern is sinusoidal. As long as the recording is sensibly linear (irradiance proportional to the final amplitude transmittance), the hologram will be a diffraction grating varying sinusoidally in transmittance, and only the first diffracted orders will be observed. One of these first-order waves will be traveling in the same direction as the object wave; this wave has been reconstructed. The other diffracted wave is traveling in a third direction and is called the conjugate wave. This is the twin wave that gave rise to so much trouble in Gabor's early scheme, since it was traveling in the same direction as both the directly transmitted (zero-order) wave and also the object wave. With the off-axis method, all three of these waves are spatially separated.

We can describe the recording of a hologram of a more complicated object with the aid of Fig. 2.2. Let O be a monochromatic wave from the object incident on the recording medium H, and let R be a wave coherent with O. The wave O contains information about the object since the object has uniquely determined the amplitude and phase of O. The object can be thin and transmitting, such as a transparency, or it can be opaque and diffusely reflecting. To make a hologram, then, we must record the incident wave field

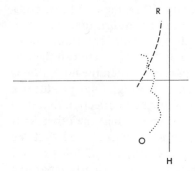

Fig. 2.2 Illustrating the recording of a general wavefront. The curves labeled O and R are schematic representations of the object and reference wavefronts, respectively.

on a photosensitive medium (usually a photographic emulsion) in such a way that the entire wave can later be reconstructed. To do this, it is necessary to record both the amplitude and phase of the wave O, which is done with the aid of the reference beam R. The total field on H is $O + R$. A square-law recording medium, such as a photographic emulsion, will respond to the irradiance of the light $|O + R|^2$. We now assume that, after processing, the hologram possesses a certain amplitude transmittance $t(x)$, which may be expressed as a function of the exposure:

$$t(x) = f[E(x)].$$

By expanding this into a Taylor's series about the average exposure E_0, and retaining only the first two terms, we can write (see Section 6.1.2 for more detail)

$$t(x) = f(E_0) + \beta E(x). \tag{2.1}$$

For our present purposes, we will simply ignore the constant term $f(E_0)$ and write the amplitude transmittance as

$$t(x) = \beta E(x) = \beta |O + R|^2$$
$$= \beta(|O|^2 + |R|^2 + OR^* + O^*R), \tag{2.2}$$

where the asterisk denotes a complex conjugate. Let us first assume that the hologram with this amplitude transmittance is illuminated with the reference wave R alone. The transmitted field at the hologram is then

$$\psi(x) = R(x) \cdot t(x) = \beta(R |O|^2 + R |R|^2 + |R|^2 O + R^2 O^*). \tag{2.3}$$

If now the reference wave R is sufficiently uniform so that $|R|^2$ is approximately constant across the hologram, then the third term of Eq. 2.3 is $\beta |R|^2 O = \text{const} \times O$. This term, then, represents a wave identical to the object wave O. This wave has all the properties of the original wave and can form an image of the object. The fact that this wave is separated from the rest can be seen most clearly by analogy with the diffraction grating hologram described above. If we consider the complex wave O of Fig. 2.2 to be composed of many plane waves of different amplitudes and directions and also consider R to be a plane wave, then each of the component plane waves interferes with R to form a grating, as in Fig. 2.1. The hologram can then be thought of as consisting of a large number of gratings. When the hologram is illuminated by the reference beam alone, each of these gratings produces a first-order diffracted plane wave of such an amplitude and direction that they all sum to the composite wave O. It can also be shown that the other first-order diffracted wave arises from the term $\beta O^{*}R^{2}$ of Eq. 2.3 and that the zero-order wave is expressed by the first two terms, $\beta R(|O|^{2} + |R|^{2})$ of Eq. 2.3. In this way, then, we see that the object wave O is clearly separated from the others and may be viewed independently.

Another way to explain this separation, due to Leith and Upatnieks, is by analogy with the idea of a carrier frequency [1]. Assume that the reference wave R is a plane wave that is incident on the hologram plane at an angle to the object wave O. This is shown in Fig. 2.3. Let the reference wave be written $R_0 e^{i\alpha x}$, where R_0 is a constant, and the linear phase shift αx indicates that the wave is incident at some angle $\varphi = \sin^{-1}(\alpha/k)$, where k is the wave number of the light, $2\pi/\lambda$. Also write the object wave as $O_0 e^{i\varphi_0}$ so that O_0 and R_0 are real numbers, and Eq. 2.2 becomes

$$t(x) = \beta[O_o^2 + R_o^2 + 2O_o R_o \cos(\alpha x - \varphi_o)]. \tag{2.4}$$

The information $O_0 e^{i\varphi_o}$ in the object wave has been transferred to a spatial carrier wave $\cos(\alpha x)$. The amplitude O amplitude modulates this carrier and the phase φ_o phase modulates the carrier. By adding the object wave to a spatial carrier frequency, we have been able to achieve the desired separation.

The degree of separation of the various terms of Eq. 2.4 depends on their

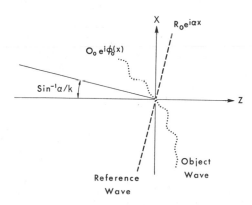

Fig. 2.3   Mathematical expressions describing the object and reference wavefronts.

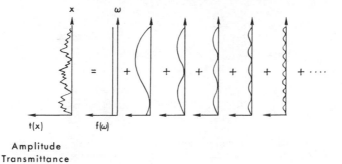

Amplitude
Transmittance

Fig. 2.4 Schematic representation of the Fourier decomposition of the transmission function $t(x)$ into a sum of elementary diffraction gratings $f(\omega)$.

spatial frequency content, which may be explained as follows. Equation 2.4 describes in one dimension the spatial variation of amplitude transmittance $t(x)$. We can think of $t(x)$ as being composed of a large number of sinusoidally varying transmission gratings, each with a different spatial frequency and phase angle (Fig. 2.4). The spatial frequencies of these composite gratings vary from 0 to some maximum radian spatial frequency ω_{max}. This is the principle of Fourier decomposition. Formally we can write

$$t(x) = \int_0^\infty f(\omega)e^{-i\omega x}\,d\omega \qquad (2.5)$$

so that $f(\omega_0)e^{-i\omega_0 x}$ describes a single grating of radian spatial frequency ω_0 and amplitude $f(\omega_0)$. When a plane wave of light is incident on the plane (Fig. 2.5), each elementary grating diffracts the light in a direction determined by the spatial frequency of the grating, as determined by the grating equation

$$\sin\theta = \frac{\lambda\omega}{2\pi} \qquad (2.6)$$

where λ is the wavelength of the incident plane wave and ω is the radian spatial frequency of the elementary grating. The superposition of all of the

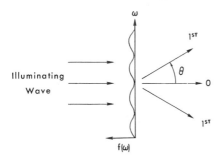

Illuminating
Wave

Fig. 2.5 Diffraction from an elementary grating.

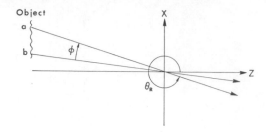

Fig. 2.6 Origin of the transmittance fluctuations due to the object alone. The interference of light from object points such as *a* and *b* produce low spatial frequency exposure fluctuations at the hologram plane.

diffracted waves yields the total field diffracted by $t(x)$. The largest angle θ (Eq. 2.6) at which any light will be diffracted depends on the largest spatial frequency $\omega_{\max}$ contained in $t(x)$. If $t(x)$ varies only slowly, then $\omega_{\max}$ will be relatively small and hence θ will remain small. If $t(x)$ varies rapidly from point to point, then $\omega_{\max}$ will be large and some light will be diffracted at large angles.

The spatial frequency content of the various terms of Eq. 2.4 depends on how each term was produced. In most cases, R_0 is independent of x, so that there will be no point-to-point variation in the transmittance. For this term, $\omega_{\max} = 0$, and no light will be diffracted away from the directly transmitted beam because of this term. On the other hand, O_o is usually a function of x, $O_o(x)$. This occurs because of self-interference between different object points, such as points *a* and *b* of Fig. 2.6. These two points will interfere so as to produce an interference pattern with radian spatial frequency

$$\omega = 2k \sin\left(\frac{\varphi}{2}\right) \cos\left(\theta_R - \frac{\varphi}{2}\right); \qquad (2.7)$$

the angles are defined in Fig. 2.6. Thus the larger the angle φ is, the greater the spatial frequency of the resultant grating will be. Hence $\omega_{\max}$ for the $O_o^2(x)$ term will depend on the angular extent of the object: the larger the angular extent of the object, the greater the spatial frequency content of $O_o(x)$. The light diffracted from the elementary gratings comprising $O_o^2(x)$ is termed *flare light*, since $\omega_{\max}$ is usually relatively small and the diffracted light caused by this term generally extends only to small angles from the directly transmitted light.

The image-producing third term of Eq. 2.4 has all of the spatial frequency components associated with $O_o(x)$ and also that of the *carrier wave*, $\cos(\alpha x)$. If α of Fig. 2.3 is made large enough, the images will be formed at large angles from the directly transmitted and flare light, since $\omega_{\max}$ for this term will be α

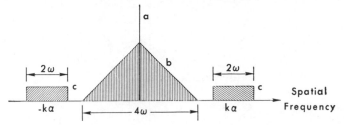

Fig. 2.7 A schematic representation of the spatial frequency spectrum of a hologram (after Leith and Upatnieks [1]).

plus the $\omega_{\max}$ of the object wave. In this way we can achieve the desired separation of the image and noise terms of Eq. 2.4. A schematic representation of the spectrum of the various terms of 2.4 is shown in Fig. 2.7.

Any nonlinearity in the recording process gives rise to higher diffracted orders, but these will generally be weak and not overlap the desired waves. The elimination of the problems of nonlinear recording is one of the major advantages of off-axis holograms.

The third term of Eq. 2.4 leads to two images; these are separated from each other, as well as from the flare light. To show this more clearly, we write Eq. 2.4 in the form

$$t(x) = \beta(O_o^2 + R_o^2 + O_oR_oe^{i(\alpha x - \varphi_o)} + O_oR_oe^{-i(\alpha x - \varphi_o)}), \tag{2.8}$$

so that the carrier wave is expressed by the term $e^{i\alpha x}$. If we now illuminate the hologram with a wave in the same direction as the reference wave, say $C(x) = C_0e^{i\alpha x}$, the third and fourth terms of Eq. 2.8 yield the two first-order waves

$$\beta C_o O_o R_o e^{2i\alpha x - i\varphi_o} \tag{2.9}$$

and

$$\beta C_o O_o R_o e^{i\varphi_o}. \tag{2.10}$$

The wave (2.10) is identical to the original object wave except for the un-important multiplicative constants. This wave is traveling in the same direction as the original object wave, but the wave described by Eq. 2.9 is traveling in a different direction. The direction of this wave is determined by the phase factor $e^{2i\alpha x}$. To see why this is so, assume for the moment that φ_0 is constant, that is, does not depend on x. Then the exponent $2\alpha x$ describes the phase of the field in the x-plane. The wavefront corresponding to such a phase in the x-plane is a *plane* wavefront tilted at an angle $\sin^{-1}(2\alpha/k)$ to the plane, as shown in Fig. 2.8a. The phase $\varphi_0(x)$ might correspond to a wavefront as shown in Fig. 2.8b, for example, so that the wavefront corresponding to the phase $2\alpha x - \varphi_0(x)$ might appear as in Fig. 2.8c. The effect of the carrier $2\alpha x$ has been to change the general direction of propagation of the wave $e^{i\varphi_0(x)}$. Hence

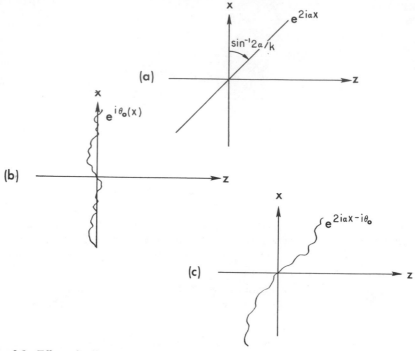

Fig. 2.8 Effect of a linear phase shift on the direction of propagation of a wave. In (a), a wave described by a phase that varies linearly with the coordinate is tilted with respect to that coordinate axis and so is propagating at an angle to it; if this linear phase shift is added to a general wave (b), the direction of propagation of this wave is altered (c).

the two waves described by Eqs. 2.9 and 2.10 are traveling in different directions and will eventually separate.

Since the phase function $\varphi_0(x)$ is *subtracted* from the carrier $2\alpha x$ in Eq. 2.9, the wave is said to be *phase conjugate* to the wave $e^{i\varphi_0(x)}$. The conjugate wave produces an image lying on the opposite side of the hologram from the illuminating source, whereas the wave $e^{i\varphi_0(x)}$ produces an image in the same location as the original object. The fact that Eq. 2.9 describes a wave that is phase conjugate to the original object wave means that the wavefront is inverted. For example, if the original object wave were a *diverging* spherical wave, Eq. 2.9 would describe a *converging* spherical wave. In the usual case the object wave is divergent, so $e^{i\varphi_0(x)}$ leads to a virtual image of the object, while $e^{-i\varphi_0(x)}$ describes a converging wave that forms a real image of the object. For this reason, the third term of Eq. 2.8 is often referred to as the real image term and the fourth term as the virtual image term. These designations, however, should not imply that these terms always lead to either a real or virtual image. There are many possible arrangements wherein they do not, so this designation should be taken as merely a matter of convenience for labeling the various terms. We will call the image resulting from the third term the *primary* image and that from the fourth term the *conjugate* image.

So far we have not mentioned much about the types of object suitable for holography. Since the original Gabor scheme of in-line holography required a strong background wave to reduce the effects of the twin image, the only suitable objects were transparencies such as opaque lettering on a clear background. Large clear areas were necessary to produce the strong coherent background. Thin transparencies were required because of the limited coherence lengths of thermal sources. With the off-axis scheme, a reference wave can be produced that is independent of the object wave (but coherent with it) so that it is possible to produce holograms of clear letters on an opaque background, for example. Such a hologram and an image formed with the reconstructed wave is shown in Figs. 2.9a and b. Note that because of the essentially specular nature of the object, one can almost tell what the object was from the hologram.

Leith and Upatnieks have demonstrated that holograms can be constructed with diffuse illumination of the object, which has some very interesting advantages. Diffusing the light that illuminates the object effectively causes each point of the object to radiate a spherical wave, so that the information concerning each point of the object is spread out over the whole hologram. Because of this, each point of the hologram now contains information about the whole object. The really striking feature of holograms made of diffuse objects is that the reconstruction can be viewed without optical aids. In viewing the reconstruction of a specularly illuminated transparency, for example, it would be possible to see only the portion of the object through which the small cone of rays entering the eye pupil had passed (Fig. 2.10a). If the transparency is diffusely illuminated, however, the viewer sees the whole object at once, since light is passing through each point of the object in all directions (Fig. 2.10b).

It is also possible, of course, to make holograms of diffusely reflecting objects. It is this type of object which gives rise to truly striking reconstructions. The object wave is viewed through the hologram as if it were a window, and the objects appear behind the window just as in the original scene. The observer may look from above, from underneath, or from any perspective allowed by the window. He may look around foreground objects to see background objects; in short he may view the complete scene from a wide range of perspectives.

Figure 2.11a is a hologram of a diffusely illuminated object scene, and Figs. 2.11b, c, and d are views of the reconstruction from three different perspectives. Note that the hologram of Fig. 2.11a is quite different in appearance from that of Fig. 2.9a. There are no discernible Fresnel diffraction

Fig. 2.9 Nondiffuse hologram. (*a*) Actual hologram of a transparency (clear bars on opaque background) that was specularly illuminated; (*b*) image formed with the hologram of (*a*).

22

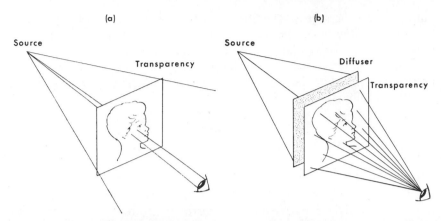

Fig. 2.10 Demonstrating the difficulty in viewing a specularly illuminated transparency as compared with a diffusely illuminated one. In (a), the only visible portion of the transparency is that small section intersected by the cone of rays entering the eye. In (b), light from all parts of the transparency reaches the eye.

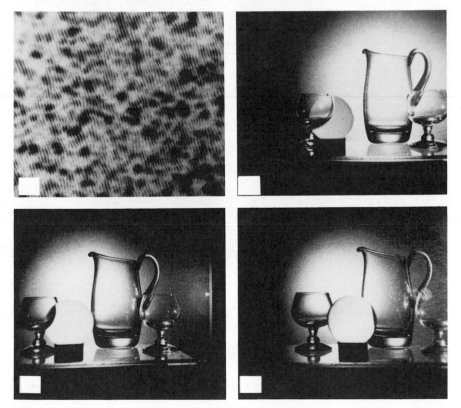

Fig. 2.11 Hologram of a diffusely illuminated object. (a) The actual hologram. (b), (c), (d) Three perspectives of the resulting image.

patterns from the object. The hologram has a granular appearance that is uniform over the whole hologram. This is the speckle pattern familiar to anyone who has viewed laser light after being reflected from or transmitted through a diffuse surface. This is the recording of the $O^2(x)$ term discussed above. Superposed on this is the fine-line, grating like structure characteristic of off-axis holograms.

2.2 ARRANGEMENTS FOR RECORDING PLANE HOLOGRAMS

Figure 2.11 illustrates the recording of a hologram and the subsequent re-construction. As shown in Fig. 2.12a, the laser beam is first expanded. This beam is then divided by means of a mirror that directs part of the beam directly onto the photographic plate, the rest of the light reflecting from the object. The hologram is the recording of the interference pattern formed by these two beams. After processing, the hologram plate may be replaced in its original position (Fig. 2.12b) and the object removed. The light that is diffracted by the hologram forms, in part, the same wavefront that was originally reflected from the object. A viewer looking through the hologram will see an undistorted view of the object, just as if it were still present.

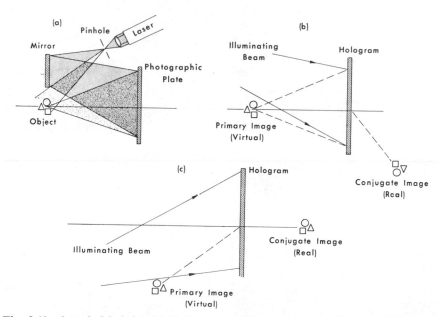

Fig. 2.12 A typical holographic arrangement. (*a*) Recording the hologram. (*b*) Recon-structing the primary object wave. (*c*) Reconstruction of an undistorted conjugate wave.

In addition to this virtual, or primary image, a real, or conjugate image will be formed on the observer's side of the hologram. This image will appear unsharp and highly distorted. However, a distortion-free real image can be formed by changing the position of the illuminating beam so that it appears to come from a mirror image of the reference beam, with respect to the hologram. In this way, an undistorted real, three-dimensional image of the object scene appears in front of the hologram as shown in Fig. 2.12c. The conjugate image suffers from a strange depth inversion, first called "pseudoscopic" by Leith et al. [2], but it was later shown by Meier [3] that this is an incomplete description of the effect. In addition to the reversed parallax of pseudoscopic imagery, there is a reversed focus of the three-dimensional optical image. It is possible to form a hologram that will produce a real image without depth inversion by using the conjugate image of another hologram as the object, and viewing the conjugate image [4]. For plane objects, such as transparencies, this effect is of course unimportant.

Holograms of transparencies may be made with an arrangement such as shown in Fig. 2.13a, and the object wave reconstructed as in Fig. 2.13b. If there is no diffuser behind the transparency, the hologram will appear as in Fig. 2.9. The image formed with such a hologram will be free from the characteristic "speckle" of diffuse laser illumination, but suffers the disadvantage that there is almost a 1:1 correspondence between object points and position on the hologram; the hologram is essentially a shadowgram of the object. Thus whereas damage to a portion of a hologram of a diffuse object has only a minor effect, damage to this hologram may result in a loss of parts of the image.

Holograms may be recorded with diverging, parallel, or converging reference beams. If care is taken to maintain the recording geometry during reconstruction, it is in fact possible to form holograms with an arbitrary reference beam; the only requirement is that it be coherent with the object beam. A hologram formed with a nearby object is called a Fresnel hologram (Fig. 2.12). There is another class of holograms that is of primary importance.

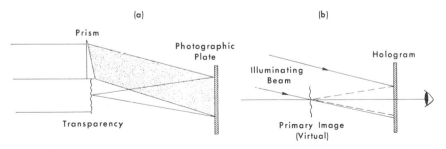

Fig. 2.13   Recording a hologram of a transparency. (a) Recording (b) Reconstructing.

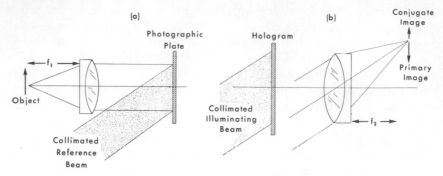

Fig. 2.14   Fourier transform hologram. (*a*) Recording. (*b*) Reconstructing.

Holograms in this class are called Fraunhofer, or Fourier transform [2] holograms, and they are recorded as shown in Fig. 2.14*a*. A plane object is placed in the focal plane of a lens so that each object point gives rise to a parallel beam of light incident on the plate. The reference beam is collimated and is incident on the plate at some angle with respect to the object beam. In this way, both the object and reference source are effectively at infinity. For

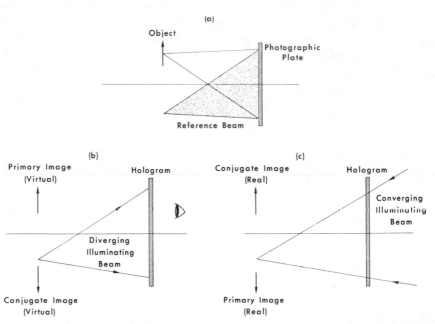

Fig. 2.15   An approximate Fourier transform hologram without lenses. (*a*) Recording. (*b*) Reconstructing.

reconstruction, a second lens is used; this lens need not be the same focal length as the lens used in recording (Fig. 2.14b). Both the primary and conjugate images will be formed in the focal plane of this lens, magnified by the ratio f_2/f_1.

Holograms of this type are sometimes called "Fourier transform" holograms, since if the photographic plate is placed a distance f_1 from the lens (Fig. 2.14), the light distribution at the hologram due to the object very closely approximates the Fourier transform of the light distribution at the object (see Appendix A).

This situation can be approximated when the object and reference point are in the same plane and no lenses are used. Such a hologram has been termed a "lensless Fourier transform" hologram [5], since to a first approximation, each object point gives rise to fringes of a single spatial frequency across the photographic plate (cf. Eq. 3.46). An arrangement for making such a hologram is shown in Fig. 2.15a. In Fig. 2.15b, a diverging illuminating beam produces both the primary and conjugate images in a plane containing the point from which the illuminating beam diverges. Both of these images are virtual. Two real images may be formed by reversing the rays in the illuminating beam as shown in Fig. 2.15c.

2.3 ARRANGEMENTS FOR RECORDING VOLUME HOLOGRAMS

A volume hologram is one for which the thickness of the recording medium is of the order of or greater than the spacing of the recorded fringes. In the Leith-Upatnieks type of off-axis hologram the basic fringe spacing depends on the angle between the object and reference beams. If this angle is greater than about 7 or 8° and the recording is made with visible light, then holograms made on a photographic film or plate must be considered volume holograms. In this case, the fringe spacing may be of the order of 2 μ while the emulsion thickness might be 5–20 μ. Therefore, virtually all holograms made on photographic film should be considered to be volume holograms.

Volume holograms have some very useful and interesting features that are distinct from the features of plane holograms. Volume holograms lend themselves very nicely to the storage of information since the whole volume of the recording medium may be utilized, as noted by van Heerden [6]. The volume hologram has been utilized by Pennington and Lin to form a three-color holographic reconstruction [7]. Denisyuk [8] introduced a new form of holography using the whole volume of the recording medium in such a way that wavefronts could be reconstructed in reflection.

These are just a few of the major useful applications involving volume holograms. In all but a few situations, an accurate analysis should include considerations of the effects of the nonzero thickness of the recording medium. Plane holograms really represent a special case in the larger class of volume holograms. Except in the few special cases of truly plane holograms, all calculations assuming a two-dimensional recording medium should be considered only as a first approximation to any real experimental situation.

The main feature of volume holograms, distinct from plane holograms, is that only one image is formed for each direction of incidence of the illuminating beam. If a volume hologram of an object is formed in the same way as for a plane hologram, an aberration-free primary image (virtual) is formed in the position of the original object only if the illuminating beam is identical with the reference beam used to record the hologram. An aberration-free conjugate image (real, depth-inverted) is formed in the position of the object only if all of the rays in the system are reversed. These situations are shown in Fig. 2.16. In Fig. 2.16a a hologram is made on a thick recording medium. In Fig. 2.16b the primary image, which is virtual, is read out by using an exact duplicate of the reference beam for the illuminating beam. In Fig. 2.16c the conjugate (real) image is formed by reversing all of the rays of Fig. 2.16a. Figure 2.17 shows why these arrangements are necessary. In 2.17a, the fringes formed by interference between light from a single object point and from the reference point are formed throughout the volume of the recording medium. For readout, as in Fig. 2.17b, the Bragg condition requires that the angle of incidence be equal to the angle of diffraction and at the same time that both of these angles satisfy the grating equation (Chapter 4). Violation of either of these conditions will limit the amount of light diffracted, and the

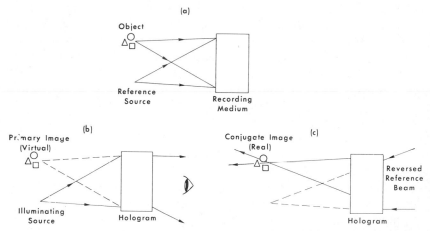

Fig. 2.16 Arrangements for recording and illuminating volume holograms. (*a*) Recording the volume hologram. (*b*) Reconstructing the primary wave. (*c*) Reconstructing the conjugate wave.

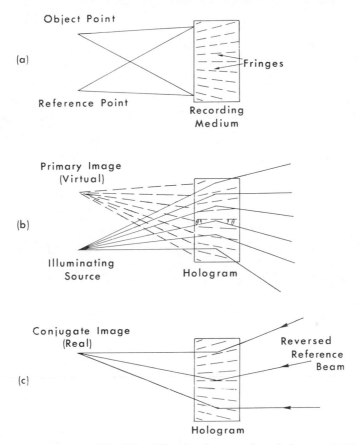

Fig. 2.17 The "Bragg condition" for diffraction from a volume hologram. (a) Formation of the fringes in the volume. (b) Reconstructing the primary wave. (c) Reconstructing the conjugate wave.

resulting image will not appear as bright as when the Bragg condition is fulfilled. If the interference fringes are considered as reflecting planes, then the Bragg condition is satisfied when the law of reflection holds and the brightest image results. Figure 2.17c shows how reversing the rays satisfies the Bragg condition. If the recording medium changes dimension during processing, such as by the shrinkage of photographic emulsion, it will generally not be possible to satisfy the Bragg condition during reconstruction, since the fringes will have changed position relative to the reference beam. Since it is no longer possible to match all of the original boundary conditions simultaneously, the resolution in the image is decreased. The image distortion caused by film shrinkage can be largely eliminated with Fraunhofer holograms. Since each

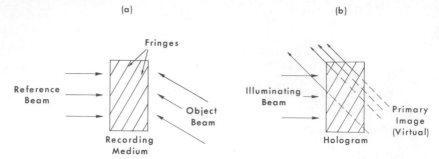

Fig. 2.18 Volume holograms that can be viewed in reflection. (*a*) Recording the hologram with object and reference waves incident in nearly opposite directions. (*b*) Reconstructing the primary wave in reflection.

object point results in a set of equally spaced, straight fringes, shrinkage results in only an increased tilt of the fringes. The Bragg condition in that case can still be satisfied for the most part by slightly changing the angle of incidence of the illuminating beam.

Volume holograms that can be viewed in reflection are recorded as shown in Fig. 2.18*a*. Here the reference beam enters the recording medium from the side opposite the object beam. The two interfering beams are traveling in approximately opposite directions and standing waves are set up in the medium. For the arrangement shown, the fringes are planes separated by approximately half the wavelength in the medium. Readout is accomplished by viewing the light reflected from these planes (Fig. 2.18*b*). Since these planes effectively form a half-wave stack interference filter, readout may be accomplished with white light, as noted by Denisyuk [8]. In this case only a small band of wavelengths from the white light source satisfy the Bragg condition at one time. The images obtained thus appear colored, and the color changes with angle of incidence. The reason for the color change with readout angle can best be understood with the aid of Fig. 2.19. Here the interference planes were recorded with wavelength λ_1 (in the medium) and so are separated by a distance $d = \lambda_1/[2 \cos (\tfrac{1}{2}\theta)]$. When the object wave is reconstructed by illuminating the hologram with a plane wave of white light as shown in Fig. 2.19*b*, significant reflection occurs only for a small band of wavelengths around λ_1. This is because the optical path difference between a typical pair of reflected (diffracted) rays, such as (1) and (2) in the figure, is given by

$$OPD = 2d \cos (\tfrac{1}{2}\theta). \tag{2.11}$$

Since $d = \lambda_1/[2 \cos (\tfrac{1}{2}\theta)]$, rays (1) and (2) are in phase and interfere constructively, so there is strong reflection in this direction for this wavelength.

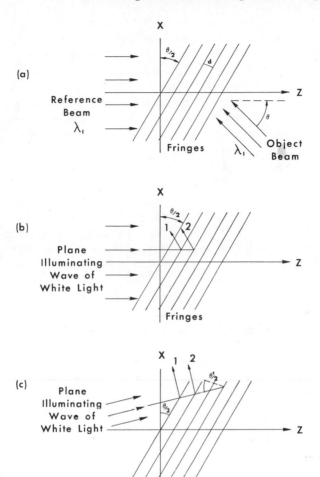

Fig. 2.19 Change in color of the reflected wave for a change in angle of incidence. (*a*) Recording the fringes. (*b*) Reconstructing the primary wave.   Changing the reflected color in accordance with the Bragg condition.

If the hologram is rotated to a viewing angle of $\frac{1}{2}\theta'$ (Fig. 2.19*c*), only the wavelength satisfying the condition

$$2d \cos\left(\tfrac{1}{2}\theta'\right) = \lambda_2 \tag{2.12}$$

or

$$\frac{\lambda_1}{\lambda_2} = \frac{\cos\left(\tfrac{1}{2}\theta\right)}{\cos\left(\tfrac{1}{2}\theta'\right)} \tag{2.13}$$

will be strongly reflected.

Color holograms may be made using this effect. A full-color object is illuminated with red, green, and blue light, and the reference beam is composed of the same combination. Each color then forms its own set of standing waves in the medium, each with different spacing. Readout can be accomplished by illuminating with white light. Each set of Bragg planes filters out the color with which it was made and a full-color reconstruction results. A more detailed analysis of color holograms, along with other possible arrangements, will be given in Chapter 7.

The arrangements described in this chapter do not indicate all that are possible, of course. We have shown only typical examples of arrangements that fall in one or the other of the two major classifications, Fresnel holograms and Fraunhofer holograms. Generally, a Fraunhofer hologram is formed when the object is a great distance away from the recording medium. A Fresnel hologram is formed when the object is near the recording medium. A more accurate mathematical distinction between these two types will be made in Chapter 3.

REFERENCES

[1] E. N. Leith and J. Upatnieks, *J. Opt. Soc. Am.*, **52**, 1123 (1962).

[2] E. N. Leith and J. Upatnieks, *J. Opt. Soc. Am.*, **54**, 1295 (1964).

[3] R. W. Meier, *J. Opt. Soc. Am.*, **56**, 219 (1966).

[4] F. B. Rotz and A. A. Friesem, *Appl. Phys. Letters*, **8**, 146 (1966).

[5] G. W. Stroke, D. Brumm, and A. Funkhouser, *J. Opt. Soc. Am.*, **55**, 1327 (1965).

[6] P. J. van Heerden, *Appl. Opt.*, **2**, 387 (1963).

[7] K. S. Pennington and L. H. Lin, *Appl. Phys. Letters*, **7**, 56 (1965).

[8] Y. N. Denisyuk, *Soviet Phys.—Doklady*, **7**, 543 (1962).

3 General Theory of Plane Holograms

3.0 INTRODUCTION

In this chapter we introduce the basic notation and coordinate systems to be used throughout the book. We then discuss the basic mathematics of making the hologram and reconstructing the wavefronts for each of the basic hologram types. In each case, we will use the fewest possible space and frequency coordinates consistent with a full understanding of the principles involved. In this way, the notation and mathematics will be kept as simple as possible. Extreme rigor will be avoided, but assumptions and their validity will be noted.

3.1 NOTATION

The origin of a Cartesian (x, y, z) coordinate system is centered in the recording medium (Fig. 3.1). Two-dimensional recording mediums (thermoplastics, thin photographic emulsions, etc.) define the $x - y$ plane. A typical object point will be denoted by (x_0, y_0, z_0). If the reference beam is derived from a point source, the coordinates of this point will be denoted by (x_R, y_R, z_R). If $z_R = \infty$, the reference beam is a plane wave and α_R will be the angle between the projection of the propagation vector $\mathbf{k}$ onto the $x - z$ plane and the z-axis. Points in the image formed with the reconstructed wavefront will be denoted by the image coordinates (x_i, y_i, z_i) (Fig. 3.1b). During reconstruction, the illuminating wave may be derived from a point source that is not identical to the reference point. We will denote the coordinates of this point by (x_c, y_c, z_c). If the reconstruction is with a plane wave, $z_c = \infty$ and we will define α_c as the angle between the projection of the $\mathbf{k}$ vector of the illuminating wave onto the $x - z$ plane with the z-axis, in the same way as with the reference wave.

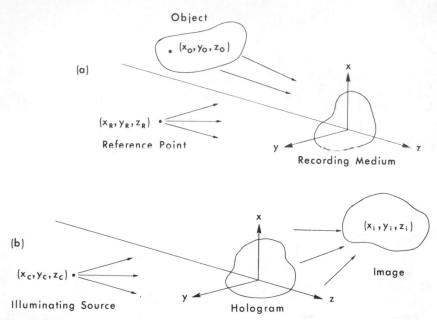

Fig. 3.1 The coordinate system and notation. (*a*) Recording the wavefront. (*b*) Reconstructing the wavefront.

The amplitude distribution of the light at the object will be denoted by $F(x_o, y_o, z_o)$. This field gives rise to a disturbance $O(x, y, z)$ at the hologram. The reference wave at the hologram can be described by $R(x, y, z)$. The time dependence of these fields has been omitted, and O and R are considered to be complex amplitudes. For reconstruction, the hologram is illuminated by a wave $C(x, y, z)$. The wave transmitted by the hologram in general gives rise to an image $G_p(x_i, y_i, z_i)$, or $G_c(x_i, y_i, z_i)$, denoting either the primary or conjugate image, respectively. Note that both object (F) and image (G) are also complex amplitudes. The final processed hologram will have a complex amplitude transmission $t(x, y, z)$ as a result of exposure with irradiance $I(x, y, z) = |H(x, y, z)|^2$, where

$$H(x, y, z) = O(x, y, z) + R(x, y, z) \qquad (3.1)$$

is the total field at the hologram. This notation is shown in Fig. 3.2*a* and *b*.

3.2   ANALYSIS

3.2.1   Fresnel Holograms

Having settled on a suitable notation, we can now proceed to describe mathematically the recording of the hologram and the subsequent wavefront

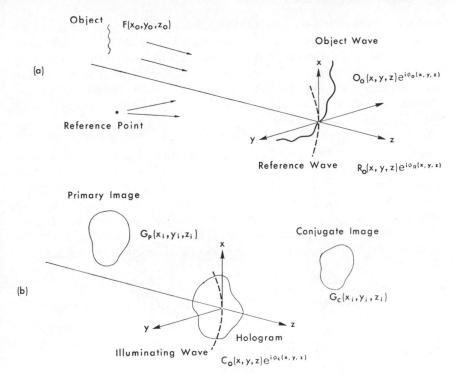

Fig. 3.2 The notation for the various optical fields encountered in holography. (*a*) Recording the wavefront. (*b*) Reconstructing the wavefront.

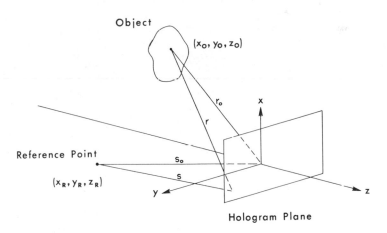

Fig. 3.3 Generalized arrangement for recording a Leith-Upatnieks off-axis, plane hologram.

reconstruction. For the discussion that follows in this chapter, the recording medium will be considered two dimensional. These will be called *plane holograms*. Surface-deformation thermoplastics and photographic emulsions that are thin compared to the spacing of the highest frequency exposure variations are good approximations to plane holograms. The case of a three-dimensional recording medium will be discussed in Chapter 4; these holograms are called *volume holograms*.

Consider the arrangement shown in Fig. 3.3. The reference point at (x_R, y_R, z_R) is the origin of a spherical reference wave. The *object* is defined by a surface S generally not including the point (x_R, y_R, z_R), no point of which lies on a line joining the reference point and any point on the recording medium. This is the general arrangement for a Leith-Upatnieks type of hologram [1]. A typical object point (x_o, y_o, z_o), lying on the surface S, gives rise to a spherical wave. The disturbance at the hologram plane caused by the object is the sum of the spherical waves from each point of the object*:

$$O(x, y, z) = \frac{-i}{2\lambda} \int\!\!\!\int_{-\infty}^{\infty} F(x_o, y_o)(1 + \cos\theta)\frac{e^{ikr}}{r}\,dx_o\,dy_o. \qquad (3.2)$$

The exponential expresses the change in phase of the wave traveling from (x_o, y_o, z_o) to (x, y, z); $1/r$ is the reduction in amplitude due to spreading of the spherical wave, and $k = 2\pi/\lambda$ is the magnitude of the wave vector, with λ the wavelength of the light.

The factor $\frac{1}{2}(1 + \cos\theta)$ is the obliquity factor, θ being the angle between the normal to the surface at (x_o, y_o, z_o) and the line joining (x_o, y_o, z_o) with (x, y, z). The factor $-i$ appears because the Huygens' wavelets are assumed advanced in phase by $\pi/2$. The two-dimensional form of Eq. 3.2 is (see Eq. A.1)

$$O(x, z) = \left(\frac{-i}{4\lambda}\right)^{1/2}\int_{-\infty}^{\infty} F(x_o)(1 + \cos\theta)\frac{e^{ikr}}{r^{1/2}}\,dx_o,$$

and this is the form we will use throughout most of this book in order to simplify many of the equations. For the purpose of classifying hologram types, however, we return to Eq. 3.2 and write the distance r as

$$r = [(x_o - x)^2 + (y_o - y)^2 + (z_o - z)^2]^{1/2}. \qquad (3.3)$$

---

* In some cases the object cannot be described on a surface, for example, for a partially transmitting object occupying some volume of space. In this case, the integral goes over to a volume integral, if all of the point scatterers in the volume are independent, or if the scattering is weak. (See [2] and [3] for a more complete treatment of Huygens' principle.)

The reference wave at the hologram plane is given by

$$R(x, y, z) = R_o \frac{e^{iks}}{s} \tag{3.4}$$

with

$$s = [(x_R - x)^2 + (y_R - y)^2 + (z_R - z)^2]^{1/2}. \tag{3.5}$$

For plane holograms, we are interested only in the plane $z = 0$, therefore

$$r = [(x_o - x)^2 + (y_o - y)^2 + z_o^2]^{1/2}. \tag{3.6}$$

In many cases in holography, $z_o^2 \sim x_o^2 + y_o^2$, and one may not use the binominal expansion for Eq. 3.6. In this case, the integral (Eq. 3.2) is extremely difficult and analysis is usually not performed explicitly. When $z_o^2 > x_o^2, y_o^2$, we may expand Eq. 3.6 as

$$r \cong z_o + \frac{x_o^2 + y_o^2}{2z_o} + \frac{x^2 + y^2}{2z_o} - \frac{xx_o}{z_o} - \frac{yy_o}{z_o} - \frac{(x_o - x)^4}{8z_o^3} - \frac{(y_o - y)^4}{8z_o^3} + \cdots \tag{3.7}$$

When we can neglect quadratic and higher order terms in x_o and y_o, we speak of *Fraunhofer holograms* (for plane reference wave); when the quadratic terms cannot be neglected, we speak of *Fresnel holograms*. We will consider Fresnel holograms first.

If we write $O(x, y) = O_o(x, y)e^{i\varphi_o(x,y)}$ and $R(x, y) = R_o(x, y)e^{i\varphi_R(x,y)}$, where O_o and R_o are real and φ_o and φ_R describe the spatial phase variation of the object and reference beams, respectively, then Eq. 3.1 becomes

$$H(x, y) = O_o e^{i\varphi_o} + R_o e^{i\varphi_R}. \tag{3.8}$$

From Eq. 2.2, this field produces an exposure

$$E(x, y) = |H(x, y)|^2, \tag{3.9}$$

which gives rise to a final amplitude transmittance

$$t(x, y) = \beta \, |H(x, y)|^2$$
$$= \beta(O_o^2 + R_o^2 + O_o R_o e^{i(\varphi_o - \varphi_R)} + O_o R_o e^{-i(\varphi_o - \varphi_R)}). \tag{3.10}$$

If we illuminate the hologram with a wave

$$C(x, y) = C_o(x, y)e^{i\varphi_c(x,y)}, \tag{3.11}$$

the transmitted wave at the hologram will be

$$\psi(x, y) = C(x, y) \cdot t(x, y)$$
$$= \beta(C_o O_o^2 e^{i\varphi_c} + C_o R_o^2 e^{i\varphi_c} + C_o O_o R_o e^{i(\varphi_c + \varphi_o - \varphi_R)} + C_o O_o R_o e^{i(\varphi_c - \varphi_o + \varphi_R)}) \tag{3.12}$$

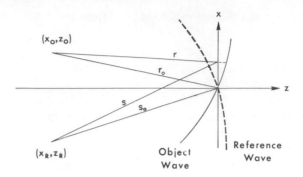

Fig. 3.4 An elementary example: recording a spherical object wave using a spherical reference wave.

This is the basic equation of holography. The way in which the phases of the various terms are expressed determines the type of hologram, either Fresnel or Fraunhofer.

In order to see more clearly how the wavefront reconstruction process works, let us analyze in detail a specific example. A suitable simple arrangement is shown in Fig. 3.4. Here we are considering a single point object and a single point reference source. Both sources emit spherical waves and so

$$O_o(x,\, y)e^{i\varphi_o(x,y)} = B\,\frac{e^{ikr}}{r} \qquad (3.13)$$

and

$$R_o(x,\, y)e^{i\varphi_R(x,y)} = A\,\frac{e^{iks}}{s} \qquad (3.14)$$

with A and B being the initial amplitude of the reference and object waves, respectively. (This result can be obtained with the use of Eq. 3.2; see, for example, [3], p. 369). We will assume that A and B are essentially constant over the area of the hologram. The distances r_0, r, s_0, and s are given by

$$r = [(x - x_o)^2 + (y - y_o)^2 + z_o^2]^{1/2}, \qquad (3.15)$$

$$r_0 = [x_o^2 + y_o^2 + z_o^2]^{1/2} \qquad (3.16)$$

and

$$s = [(x - x_R)^2 + (y - y_R)^2 + z_R^2]^{1/2}, \qquad (3.17)$$

$$s_0 = (x_R^2 + y_R^2 + z_R^2)^{1/2}. \qquad (3.18)$$

From Eqs. 3.13 and 3.15 we have

$$\varphi_o(x,\, y) = k(r - r_o)$$

$$= \frac{2\pi}{\lambda}\{[(x - x_o)^2 + (y - y_o)^2 + z_o^2]^{1/2} - (x_o^2 + y_o^2 + z_o)^{1/2}\}, \qquad (3.19)$$

which is the phase of the object wave in the hologram plane relative to the origin. Expansion of the square roots yields

$$\varphi_o(x, y) = \frac{2\pi}{\lambda}\left[\frac{1}{2z_o}(x^2 + y^2 - 2xx_o - 2yy_o) + \cdots\right]. \qquad (3.20)$$

The terms omitted in Eq. 3.20 contain $(1/z_0)^3$, $(1/z_0)^5$, and so on. For our present purposes we are interested only in the first-order terms; the higher-order terms in some cases are not negligible, however, and they give rise to aberrations of the reconstructed wave. These higher-order terms will be discussed in Chapter 5.

A similar expression is obtained for the phase of the reference wave:

$$\varphi_R(x, y) = \frac{2\pi}{\lambda}\left[\frac{1}{2z_R}(x^2 + y^2 - 2xx_R - 2yy_R) + \cdots\right]. \qquad (3.21)$$

Similarly, for the illuminating wave,

$$\varphi_c(x, y) = \frac{2\pi}{\lambda}\left[\frac{1}{2z_c}(x^2 + y^2 - 2xx_c - 2yy_c) + \cdots\right]. \qquad (3.22)$$

In the third term of Eq. 3.12, these phases add in the combination

$$\varphi_o - \varphi_R + \varphi_c = \frac{\pi}{\lambda}\left[(x^2 + y^2)\left(\frac{1}{z_o} - \frac{1}{z_R} + \frac{1}{z_c}\right)\right.$$
$$\left. - 2x\left(\frac{x_o}{z_o} - \frac{x_R}{z_R} + \frac{x_c}{z_c}\right) - 2y\left(\frac{y_o}{z_o} - \frac{y_R}{z_R} + \frac{y_c}{z_c}\right)\right]. \qquad (3.23)$$

Following Meier [4], we now consider this term as representing the first-order term of the reconstructed spherical wave,

$$\Phi_p^{(12} = \frac{\pi}{\lambda}\left(\frac{x^2 + y^2 - 2x\mathbf{X}_p - 2y\mathbf{Y}_p}{\mathbf{Z}_p}\right) \qquad (3.24)$$

with $\mathbf{Z}_p$, $\mathbf{X}_p$, $\mathbf{Y}_p$ the center coordinates of the spherical wave. The subscript p indicates that we are referring to the primary wave. From Eqs. 3.23 and 3.24,

$$\mathbf{Z}_p = \frac{z_o z_R z_c}{z_R z_c - z_o z_c + z_o z_R}$$

$$\mathbf{X}_p = \frac{x_o z_R z_c - x_R z_o z_c + x_c z_o z_R}{z_R z_c - z_o z_R + z_o z_R} \qquad (3.25)$$

$$\mathbf{Y}_p = \frac{y_o z_R z_c - y_R z_o z_c + y_c z_o z_R}{z_R z_c - z_o z_c + z_o z_R}.$$

If the illuminating wave is identical with the reference wave, then $z_c = z_R$, $x_c = x_R$, $y_c = y_R$, and

$$\mathbf{Z}_p = z_o, \qquad \mathbf{X}_p = x_o, \qquad \mathbf{Y}_p = y_o \qquad (3.26)$$

so that the reconstructed wave exactly corresponds to the original object wave.

Performing the same set of operations for the fourth term of Eq. 3.12 yields

$$\mathbf{Z}_c = \frac{z_o z_R z_c}{z_o z_R - z_c z_R + z_c z_o}$$

$$\mathbf{X}_c = \frac{x_c z_o z_R - x_o z_c z_R + x_R z_c z_o}{z_o z_R - z_c z_R + z_c z_o} \qquad (3.27)$$

$$\mathbf{Y}_c = \frac{y_c z_o z_R - y_o z_c z_R + y_R z_c z_o}{z_o z_R - z_c z_R + z_c z_o}$$

where the subscript c means we are referring to the conjugate wave. Again, for the illuminating wave identical with the reference wave, $z_R = z_c$, $x_R = x_c$, $y_R = y_c$, then

$$\mathbf{Z}_c = \frac{z_o z_R}{2z_o - z_R}$$

$$\mathbf{X}_c = \frac{2x_R z_o - x_o z_R}{2z_o - z_R} \qquad (3.28)$$

$$\mathbf{Y}_c = \frac{2y_R z_o - y_o z_R}{2z_o - z_R}$$

so that this wave yields an image on the opposite side of the hologram plane from the object, displaced from the axis by an amount that depends on the original object and reference point coordinates.

Whether the images are real or virtual depends on the sign of $\mathbf{Z}_p$ and $\mathbf{Z}_c$. Since z_o is always negative (because of the way we have set up the coordinate system) a negative $\mathbf{Z}_p$ or $\mathbf{Z}_c$ means a virtual image. Since z_R and z_c are arbitrary, both images may be real or virtual.

When the illuminating wave is not identical to the reference wave, the first-order image positions are given by Eqs. 3.25 and 3.27. We shall see later that this gives rise to magnification along with aberrations. Magnification and aberrations also occur when the illuminating wavelength is not the same as the wavelength of the reference beam.

One very common arrangement used in holography involves a plane reference wave. In this case $z_R = -\infty$ and we denote the direction of the wave by α_R, the angle between the propagation vector $\mathbf{k}$ and the z-axis in the x-z plane. Analogously, we reconstruct with a plane wave ($z_c = -\infty$) at an

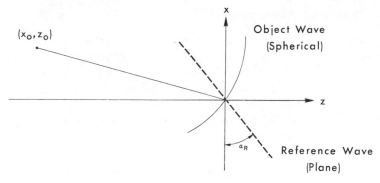

Fig. 3.5 Recording a spherical object wave using a plane reference wave.

angle α_c. This case is shown in Fig. 3.5. In this situation we have

$$\varphi_R = \frac{2\pi}{\lambda} x \sin \alpha_R$$
$$\varphi_c = \frac{2\pi}{\lambda} x \sin \alpha_c$$

(3.29)

so that

$$\varphi_o - \varphi_R + \varphi_c = \frac{\pi}{\lambda}\left\{\frac{x^2 + y^2}{z_o} - 2x\left[\frac{x_o}{z_o} - \sin \alpha_R + \sin \alpha_c\right] - 2y\frac{y_o}{z_o}\right\}$$

(3.30)

and

$$\mathbf{Z}_p = z_o$$
$$\mathbf{X}_p = x_o - z_o \sin \alpha_R + z_o \sin \alpha_c$$
$$\mathbf{Y}_p = y_o.$$

(3.31)

Again, for the illuminating beam identical with the reference beam ($\sin \alpha_R = \sin \alpha_c$), one of the reconstructed waves exactly corresponds to the original object wave. For the conjugate image we have

$$\varphi_R - \varphi_o + \varphi_c = \frac{2\pi}{\lambda}\left\{x \sin \alpha_R - \frac{1}{2z_o}(x^2 + y^2 - 2xx_o - 2yy_o) + x \sin \alpha_c\right\}$$

(3.32)

and

$$\mathbf{Z}_c = -z_o$$
$$\mathbf{X}_c = x_o - z_o \sin \alpha_R - z_o \sin \alpha_c$$
$$\mathbf{Y}_c = y_o.$$

(3.33)

Therefore both images are displaced from the axis and from each other by amounts that depend on the offset angle of reference and illuminating beams.

3.2.2 Fraunhofer Holograms

The designation *Fresnel hologram* has been applied to the case where the terms of order greater than two in x_o and y_o may be neglected in the expansion (Eq. 3.7). Generally speaking, a Fresnel hologram is formed when the object is near the hologram plane. When the object size is small compared to its distance z_o from the hologram plane, we need consider only the first-order terms in x_o and y_o in Eq. 3.7. In this case, one speaks of a Fraunhofer hologram if the reference wave is plane. The obliquity factor of Eq. 3.2 is nearly unity and

$$r \sim z_o + \frac{x^2 + y^2}{2z_o} - \frac{xx_o}{z_o} - \frac{yy_o}{z_o}. \tag{3.34}$$

The denominator r in the integral of Eq. 3.2 may be written as z_o and taken outside the integral; Eq. 3.2 therefore becomes

$$O(x, y) = \frac{-i}{2\lambda z_o} e^{ikz_o} \exp\left(ik\frac{(x^2 + y^2)}{2z_o}\right)$$

$$\times \iint_{-\infty}^{\infty} F(x_o, y_o) \exp\left(-\frac{ik}{z_o}(xx_o + yy_o)\right) dx_o\, dy_o \tag{3.35}$$

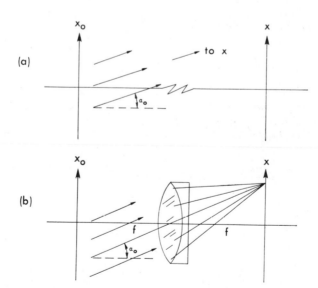

Fig. 3.6 Imaging the far field with a lens. (*a*) Desired far-field distribution. (*b*) How the desired field distribution forms in the focal plane of a lens.

for a *plane* hologram. Strictly speaking, we can neglect the second- and higher-order terms in x and y only in the limiting case $z_o \to \infty$. However, Eq. 3.35 is valid for the situation where a large, well-corrected lens is placed a focal length away from the object and the x, y plane is in the rear focal plane of the lens. To understand why this is true, consider the two situations shown in Fig. 3.6. In Fig. 3.6a, we see that the disturbance in the x, y plane, a great distance from the object, can be considered as arising from the superposition of plane waves originating from each point of the object and traveling in the direction defined by the angle α_o. However, if a well-corrected lens is placed a focal length away from the object (Fig. 3.6b), all of the light from the object traveling at an angle α_o will come to a focus at a point x in the focal plane of the lens. Since the optical path from the wavefront of the wave traveling at an angle α_o to x is the same for all rays, one obtains essentially the same interference effects predicted by Eq. 3.35. The reason for placing the lens a distance f (one focal length) from the object is discussed in Appendix A.

Hence we can write Eq. 3.35 as

$$O(x, y) = C \int\!\!\int_{-\infty}^{\infty} F(x_o, y_o) \exp\left(-i\frac{k}{f}(xx_o + yy_o) \right) dx_o\, dy_o \qquad (3.36)$$

where $C = (-i/2z_o\lambda)e^{ikz_o}$. By suitably defining the object function $F(x_o, y_o)$ so that it goes to zero beyond the object, the limits on the integral can be extended to $\pm\infty$. In this case we see that, apart from a constant, the amplitude distribution of the light at the hologram plane from the object is very nearly the two-dimensional Fourier transform of the object distribution (see Appendix A). This same result is obtained, of course, when z_o is very large compared with $(x^2 + y^2)^{1/2}$. Holograms of this sort are often called Fourier transform holograms, when the reference wave is plane.

This type of hologram has some very interesting properties. Suppose the object consists of a single point at $x_o = a$ (considering only the two-dimensional example as in Fig. 3.7). Then $F(x_o, y_o) = \delta(x_o - a)$ and

$$O(x) = \text{constant} \times e^{-i(k/f)ax}. \qquad (3.37)$$

This represents a plane wave incident on the hologram plane at an angle $\alpha_o = \sin^{-1}(a/f)$.

If now the reference wave is a plane wave striking the hologram plane at an angle α_R, we have

$$\delta(x) \equiv \varphi_R(x) - \varphi_o(x) = kx(\sin \alpha_R - \sin \alpha_o) \qquad (3.38)$$

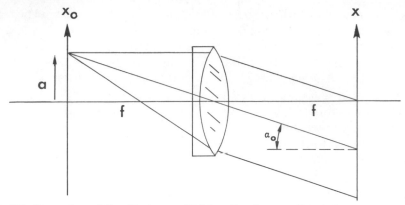

Fig. 3.7 Formation of the object wave $O(x)$ in a Fourier transform hologram system. A single object point is located at $x_o = a$.

for the phase difference between the two waves. There will be a bright fringe for $\delta = 2m\pi$ where m is an integer, or for

$$x = \frac{m(2\pi/k)}{\sin \alpha_R - \sin \alpha_o} \qquad (3.39)$$

where $k = 2\pi/\lambda$. Thus a single object point in a Fourier transform system forms a set of straight fringes of spacing

$$\Delta x_f = \Delta m \frac{\lambda}{\sin \alpha_R - \sin \alpha_o} = \frac{\lambda}{\sin \alpha_R - \sin \alpha_o}.$$

The inverse of this fringe spacing, $1/\Delta x_f$, is the basic spatial frequency recorded on the hologram. Figure 3.8 shows how the spatial frequency ν_s varies as a function of the total angle φ between the two beams for two common situations. In (1), the normal to the recording plane bisects the angle formed by the two beams so that $\alpha_o = -\alpha_R$; in (2), one of the beams is incident normally, say $\alpha_o = 0$. The wavelength used for the computation is the He-Ne laser wavelength, 0.633μ. If the hologram is a Fresnel hologram, there are of course many values of α_o—in this case the curves of Fig. 3.8 are still useful if we call α_o the angle subtended at the hologram plane by the center of the object—ν_s will then be the average spatial frequency of the hologram.

Returning to the problem at hand, we have, as usual, the total field at the hologram given by (in one dimension)

$$H(x) = O_o(x)e^{i\varphi_o(x)} + R_o(x)e^{i\varphi_R(x)} \qquad (3.40)$$

which can be written

$$H(x) = O_o e^{-ikx \sin \alpha_o} + R_o e^{-ikx \sin \alpha_R}, \qquad (3.41)$$

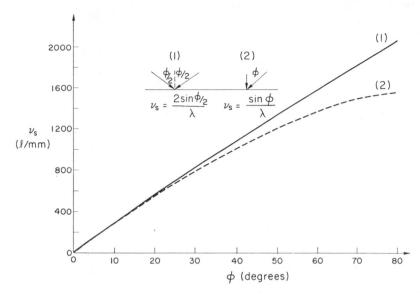

Fig. 3.8 The spatial frequency of the interference fringes ν_s as a function of the total angular separation φ between the object and reference beams for two common configurations.

so that the exposure is (by Eq. 3.9)

$$E(x) = (O_o^2 + R_o^2 + O_oR_oe^{-ikx(\sin \alpha_o - \sin \alpha_R)} + O_oR_oe^{-ikx(\sin \alpha_R - \sin \alpha_o)}). \quad (3.42)$$

We will assume the amplitude transmittance to be given as in Eq. 2.2: $t(x) = \beta E(x)$. If we illuminate the hologram with a plane wave

$$C(x) = C_o(x)e^{i\varphi_c(x)} = C_oe^{-ikx \sin \alpha_c}, \quad (3.43)$$

then the transmitted wave is of the form

$$\psi(x) = \beta(C_oO_o^2e^{-ikx \sin \alpha_c} + C_oR_o^2e^{-ikx \sin \alpha_c}$$
$$+ C_oO_oR_oe^{-ikx(\sin \alpha_o - \sin \alpha_R + \sin \alpha_c)}$$
$$+ C_oO_oR_oe^{-ikx(\sin \alpha_R - \sin \alpha_o + \sin \alpha_c)}). \quad (3.44)$$

Each of these terms represents a plane wave, since the phase advances linearly with x. The two waves represented by the first two terms are traveling in the same direction as the illuminating wave. This is the zero-order wave from the gratinglike pattern on the hologram. The third and fourth terms are equal amplitude plane waves traveling in two different directions that are symmetrical about the zero order. These represent the two first orders of the grating, and they both image the object point at infinity. If $\alpha_c = \alpha_R$, then we see that the third term represents the reconstructed object wave. If the recording is linear, as has been assumed, there are no higher diffracted orders

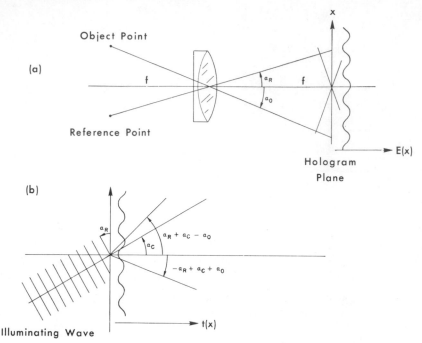

Fig. 3.9 The Fourier transform hologram. (a) The wavefront from a single object point is recorded as a grating that varies sinusoidally in transmittance. (b) The object wave is reconstructed as one of the first-order diffractions from the hologram.

since the grating transmittance is sinusoidal. This can be seen from Eq. 3.42, which can be written

$$E(x) = \{O_o^2 + R_o^2 + 2O_oR_o \cos [kx(\sin \alpha_o - \sin \alpha_R)]\}. \qquad (3.45)$$

Figure 3.9 shows schematically the recording and reconstructing of a single-point Fourier transform hologram.

The extension to more than a single object point is simple, if we extend the diffraction grating analogy. Each object point yields a plane wave at the hologram plane which interferes with a plane reference wave, producing a multiplicity of coherently superposed sinusoidal gratings. Each grating then forms two diffracted first orders, which form the two images at infinity of each object point. The modulation, or fringe visibility, of each of the component gratings decreases as the number of object points increases, but as we will see in Chapter 8 when discussing information storage capabilities, the allowable object size (or number of object points) becomes quite large before the modulation decreases to a point where no hologram can be recorded.

There is one other special recording arrangement for producing Fourier transform holograms that deserves mention here. If the reference beam is

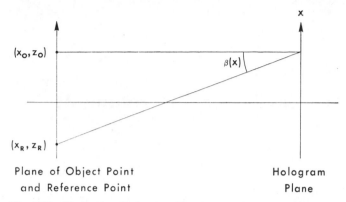

Fig. 3.10   Illustrating the lensless Fourier transform hologram.

derived from a point source that is in the same plane as the object, and no lenses are used, then the sphericities of the two spherical waves at the holo-gram plane (one from an object point, the other from the reference point) tend to cancel. Thus an approximately constant-spatial-frequency grating is produced for each object point, and we have a "lensless Fourier transform hologram" [5]. To see how this comes about, consider the arrangement of Fig. 3.10. For an object point at x_o and the reference point at x_R, the angle between the two rays at the hologram is $\beta(x)$, where

$$\beta(x) = \tan^{-1}\left(\frac{x_o - x}{z_o}\right) + \tan^{-1}\left(\frac{x - x_R}{z_o}\right)$$

$$= \frac{x_o - x_R}{z_o} - \frac{1}{3z_o^3}[x_o^3 - x_R^3 - 3x(x_o^2 - x_R^2) + 3x^2(x_o - x_R)] + \cdots,$$

$$(3.46)$$

which does not vary strongly with x for $x/z_o \ll 1$. Since the angle between the two interfering beams is approximately constant across the hologram, the spacing of the fringes is constant as well. We therefore have an approximate Fourier transform hologram, although not in the sense that the field distribu-tion at the hologram due to the object is the Fourier transform of the object distribution.

REFERENCES

[1] E. N. Leith and J. Upatnieks, *J. Opt. Soc. Am.*, **52**, 1123 (1962).

[2] B. B. Baker and E. T. Copson, *The Mathematical Theory of Huygens' Principle*, Oxford University Press, London, 1950.

[3] M. Born and E. Wolf, *Principles of Optics*, Pergamon Press Ltd., Oxford, 1964.

[4] R. W. Meier, *J. Opt. Soc. Am.*, **55**, 987 (1965).

[5] G. W. Stroke, D. Brumm, and O. Funkhouser, *J. Opt. Soc. Am.*, **55**, 1327 (1965).

4   General Theory of Volume Holograms

4.0   INTRODUCTION

Volume holograms, as defined in Chapter 3, are those for which the thickness of the recording medium is not negligible compared to the fringe spacing. In the Leith-Upatnieks off-axis type of hologram, with the images well separated in space, the information recorded on the hologram is contained in a spatial frequency band centered on the spatial carrier frequency determined by the offset angle (see Fig. 3.8). It is the fringe spacing associated with this carrier frequency which determines whether a hologram is a plane hologram or a volume hologram. If the center of the object is angularly separated from the reference point by an angle φ, then the fringe spacing of the carrier frequency is of the order of

$$d \sim \frac{\lambda}{\sin \varphi} \tag{4.1}$$

where λ is the wavelength of the recording light. When d is of the order of, or smaller than, the thickness of the recording medium, the hologram must be considered a volume hologram. By far the most common material used for recording is photographic film with emulsion thicknesses ranging from a few microns to $20 \, \mu$ and more. Therefore, most holograms recorded on film should be considered volume holograms. In the gray area of low carrier frequency holograms on thin emulsions, most of the equations for plane holograms are valid enough, but a careful analysis of any experiment should include consideration of the nonzero emulsion thickness. Owing to the greatly increased difficulty of solving the three-dimensional problem, most analyses to date have considered only the two-dimensional problem. For many experiments, these can be considered as reasonable first approximations. For surface deformation thermoplastics and relief image holograms, the two-dimensional solutions are, of course, accurate.

4.1 FRINGES IN THREE DIMENSIONS

To begin our analysis of volume holograms, we note that the fringes formed by the interference of the object beam and the reference beam are located throughout the depth of the recording medium. The direction and spacing of the fringes depend on the location and angular separation of the object and reference point. In general, for a single reference point, each object point gives rise to fringes in the recording medium; the spacing and direction of these fringes vary with position. For object and reference point both at infinity, but angularly separated, the fringe spacing and direction will be constant. Since an arbitrary object wave may be decomposed into a spectrum of plane waves, we will consider only the simple problem of finding the spacing and orientation of the fringes caused by the interference of two plane waves incident on a three-dimensional recording medium.

Consider the situation shown in Fig. 4.1. Here we are assuming that the **k** vectors of the two interfering beams lie in the x-z plane. The coordinate system is centered in the recording medium, and all directions refer to directions in the medium. The associated directions outside the medium may be found by use of Snell's law. Let the two interfering waves be denoted by u_o and u_R. The directions of propagation of these two waves in the medium make angles θ_o and θ_R with the z-axis, respectively; the positive direction is as shown. Let

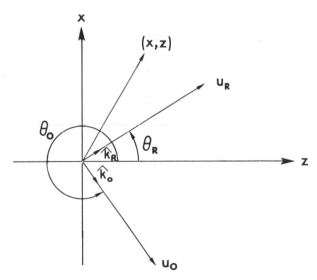

Fig. 4.1 Defining the notation to be used for the analysis of three-dimensional fringes.

$k' = 2\pi/\lambda'$ (with λ' the wavelength in the medium) be the common wave number of the two waves, and $\hat{k}_o$ and $\hat{k}_R$ be unit vectors in the direction of propagation so that

$$k'_o \equiv \hat{k}_o k'$$
$$k'_R \equiv \hat{k}_R k' \tag{4.2}$$

are the wave vectors of the two waves. If $\mathbf{r}$ is the vector position of a point (x, z) in the medium, then we may write

$$u_o = a_o e^{i k'_o \cdot \mathbf{r}} = a_o e^{i\varphi_o(x,z)}$$
$$u_R = a_R e^{i k'_R \cdot \mathbf{r}} = a_R e^{i\varphi_R(x,z)}. \tag{4.3}$$

Now if we define $\hat{x}$ and $\hat{z}$ as unit vectors along the x and z axes, respectively, then

$$\mathbf{r} = \hat{x}x + \hat{z}z \tag{4.4}$$

and so

$$\hat{k}_o = \hat{z} \cos \theta_o + \hat{x} \sin \theta_o$$
$$\hat{k}_R = \hat{z} \cos \theta_R + \hat{x} \sin \theta_R. \tag{4.5}$$

Thus we find for the phase of the wave u_o,

$$\varphi_o(x, z) = k'_o \cdot \mathbf{r} = k'z \cos \theta_o + k'x \sin \theta_o \tag{4.6}$$

and similarly for u_R,

$$\varphi_R(x, z) = k'_R \cdot \mathbf{r} = k'z \cos \theta_R + k'x \sin \theta_R. \tag{4.7}$$

The phase difference between the two waves as a function of position is given by

$$\delta(x, z) = \varphi_R(x, z) - \varphi_o(x, z)$$
$$= k'z(\cos \theta_R - \cos \theta_o) + k'x(\sin \theta_R - \sin \theta_o) \tag{4.8}$$

A bright fringe is formed along the locus of all points for which $\delta(x, z) = 2m\pi$ with m an integer, or for

$$k'z(\cos \theta_R - \cos \theta_o) + k'x(\sin \theta_R - \sin \theta_o) = 2m\pi. \tag{4.9}$$

Solving this for x, we obtain the equation defining the position of the fringes:

$$x = \left[\frac{\cos \theta_o - \cos \theta_R}{\sin \theta_R - \sin \theta_o} \right] z + \frac{m\lambda'}{\sin \theta_R - \sin \theta_o}. \tag{4.10}$$

The coefficient of z defines the slope of the straight fringe, and the second term on the right is the x-intercept. The integer m is the order of interference, which in general will depend on the phase difference between u_o and u_R at the point $(x, z) = (0, 0)$. The fringe spacing in the x-direction is the change in x corresponding to a change in m of one:

$$\Delta x = \frac{\lambda'}{\sin \theta_R - \sin \theta_o} \cdot \Delta m = \frac{\lambda'}{\sin \theta_R - \sin \theta_o} = d. \qquad (4.11)$$

The corresponding spatial frequency is

$$\nu_s = \frac{1}{d} = \frac{\sin \theta_R - \sin \theta_o}{\lambda'} \qquad (4.12)$$

This is the spatial frequency corresponding to a plane hologram. The radian spatial frequency in the x-direction is

$$\omega = 2\pi\nu_s = k'(\sin \theta_R - \sin \theta_o). \qquad (4.13)$$

In order to express ω as a function of the total angular separation between the two waves, define

$$\varphi \equiv 2\pi - \theta_o + \theta_R \qquad (4.14)$$

so that

$$\omega = k'(\sin \theta_R - \sin \theta_o)$$
$$= 2k' \sin \frac{\varphi}{2} \cos \left(\theta_R - \frac{\varphi}{2} \right). \qquad (4.15)$$

If α is the angle that the fringes make with the z-axis (see Fig. 4.2), then

$$\tan \alpha = \frac{\cos \theta_o - \cos \theta_R}{\sin \theta_R - \sin \theta_o} = \tan \left(\theta_R - \frac{\varphi}{2} \right) \qquad (4.16)$$

so that

$$\alpha = \theta_R - \frac{\varphi}{2} \qquad (4.17)$$

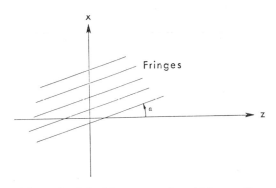

Fig. 4.2 Schematic of the fringe system in a thick recording medium.

and

$$\omega = 2k' \sin \frac{\varphi}{2} \cos \alpha. \tag{4.18}$$

This is the radian spatial frequency recorded in the x-direction for fringes produced by interference between two plane waves whose directions of propagation differ by an angle φ.

4.2 DIFFRACTION FROM A THREE-DIMENSIONAL GRATING

4.2.1 The Three-Dimensional Grating Equation

Now that the location and orientation of the fringes in the recording medium are known, we need to know how light is diffracted from such a three-dimensional grating. To do this, we begin with an approximation to the Fresnel-Kirchhoff diffraction integral. Referring to Fig. 4.3, we suppose that there is a line source at P_c which extends in the y-direction. The diffracting aperture lies in the x-y plane and we are assuming no y-variation; therefore the final integral will be two dimensional instead of three dimensional. The field at some point P_i located to the right of the aperture is given by

$$U(P_i) = A \cos \delta \left(\frac{-i}{\lambda} \right)^{1/2} \int_{\text{aperture}} \frac{e^{ik(r+s)}}{\sqrt{r+s}} \, dx. \tag{4.19}$$

The quantity A is the amplitude of the cylindrical wave at unit distance from the line source P_c. If Q is a point in the aperture, then the distances r and s are $\overline{P_cQ}$ and $\overline{QP_i}$, respectively. The angle δ is the angle between the line P_cP_i and the z-axis. If the source P_c is located at (x_c, z_c) and P_i at (x_i, z_i) then

$$\begin{aligned} r_o^2 &= x_c^2 + z_c^2 \\ s_o^2 &= x_i^2 + z_i^2; \end{aligned} \tag{4.20}$$

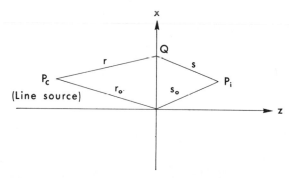

Fig. 4.3 The notation for the integral (4.19).

and if we assume that all distances are large compared with the extent of the aperture, we may write

$$r \sim r_o - \frac{x_c x}{r_o} + \frac{x^2}{2r_o} + \cdots$$

$$s \sim s_o - \frac{x_i x}{s_o} + \frac{x^2}{2s_o} + \cdots$$

(4.21)

by using the binomial expansion and ignoring the higher-order terms. With this approximation, Eq. 4.19 becomes

$$U(P_i) = A \cos \delta \left(\frac{-i}{\gamma r_o s_o} \right)^{1/2} e^{ik(r_o + s_o)}$$

$$\times \int_{aperture} \exp\left[-ik\left(\frac{x x_c}{r_o} - \frac{x^2}{2r_o} + \frac{x x_i}{s_o} - \frac{x^2}{2s_o} \right) \right] dx, \quad (4.22)$$

where we have made the approximations $(rs)^{-1/2} \sim (r_o s_o)^{-1/2}$. Now define the direction cosines of the rays from P_c to O and from O to P_i as (Fig. 4.4)

$$l_c = -\frac{z_c}{r_o} \qquad l_i = \frac{z_i}{s_o}$$

$$m_c = -\frac{x_c}{r_o} \qquad m_i = \frac{x_i}{s_o}$$

(4.23)

so that

$$U(P_i) = A \cos \delta \left[\frac{-i}{\lambda r_o s_o} \right]^{1/2} e^{ik(r_o + s_o)}$$

$$\times \int_{aperture} \exp\left\{ -ik\left[x(m_i - m_c) - \frac{x^2}{2r_o} - \frac{x^2}{2s_o} \right] \right\} dx. \quad (4.24)$$

As the distances r_o, $s_o \to \infty$, we may neglect the quadratic terms in the exponential under the integral. We assume that the factor A outside the integral tends to infinity along with r_o, s_o so that $U(P_i)$ does not vanish. This is the simpler case of Fraunhofer diffraction and we will consider only this case.

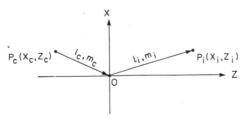

Fig. 4.4 Defining the direction cosines of the illuminating and diffracted waves.

Hence we consider the case for $r_o, s_o \to \infty$, assuming that the terms in front of the integral (Eq. 4.24) approach a constant C in the limit. We thus have

$$U(m_i) = C \int_{\text{aperture}} e^{-ikx(m_i-m_c)} \, dx \qquad (4.25)$$

as the basic equation governing Fraunhofer diffraction from a *thin* aperture.

To pass to the three-dimensional case, we assume that the integral (Eq. 4.25) yields the diffracted field in the direction m_i because of an elementary diffracting aperture of thickness dz located at $z = 0$. This is the usual "thin aperture" case. For an aperture with a nonnegligible thickness, we assume that we can just sum the contributions from each elementary thickness dz. This requires the assumption that the field diffracted from each dz is weak, so that the wave incident on each elementary aperture is the same. The solution is then only a first-order approximation for the case of weak diffraction. Burckhardt [1] has found numerical solutions for the more rigorous case. Leith et al. [2] and Ramberg [3] have essentially solved the weak diffraction problem. For a thick aperture, then, the contribution to the field in the direction $(l_i m_i)$ due to an elementary plane aperture dz at z is

$$d[U(l_i, m_i)] = \left[C \int_{\text{aperture}} e^{-ikx(m_i-m_c)} e^{-ikz(l_i-l_c)} \, dx \right] dz. \qquad (4.26)$$

The geometry is shown in Fig. 4.5. The recording medium has a thickness t.

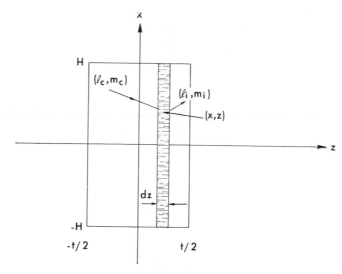

Fig. 4.5 Diffraction at a "thick aperture" as an extension of the usual "thin aperture" case.

Because of the interference fringes within the aperture, the actual diffraction problem is not that of a clear aperture, but an aperture in which there is a spatial transmittance variation $G(x, z)$, called the pupil function. Introducing the pupil function and summing all of the elementary contributions (Eq. 4.26) over an aperture $2H$ and thickness t, we obtain

$$U(l_i, m_i) = C \int_{-H}^{H} \int_{-t/2}^{t/2} G(x, z) e^{-ikx(m_i - m_o)} e^{-ikz(l_i - l_c)} \, dx \, dz. \qquad (4.27)$$

4.2.2 The Pupil Function $G(x, z)$

Before we can evaluate the integral (Eq. 4.27), we must find the form of the pupil function $G(x, z)$. To do this, we assume that the recorded signal of interest is proportional to the incident irradiance. Since the recorded signal of interest is actually the amplitude transmittance $t(x)$, the actual form is $t(x, z) = f(E_o) + \beta |H(x, z)|^2$, but the inclusion of the constant term $f(E_o)$ just adds unnecessary complication at this point, so we will omit it. The irradiance produced by the interference of the two waves of Fig. 4.1 is

$$|H(x, z)|^2 = (u_o + u_R)(u_o^* + u_R^*) = a_o^2 + a_R^2 + 2a_o a_R \cos [\delta(x, z)], \quad (4.28)$$

where the phase difference function $\delta(x, z)$ is given by Eq. 4.8:

$$\delta(x, z) = k'z(\cos \theta_R - \cos \theta_o) + k'x(\sin \theta_R - \sin \theta_o).$$

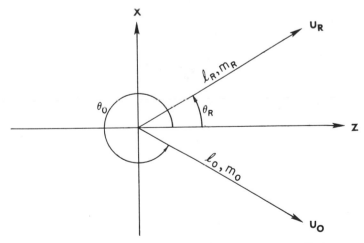

Fig. 4.6 Defining the direction cosines of object and reference waves. Both are assumed to be plane waves in this analysis.

Note that we are allowing for the possibility of different recording and illuminating wavelengths: the k' of Eq. 4.8 is the wave number of the recording light in the medium, and k of Eq. 4.27 is the wave number of the illuminating light in the medium. We now define two more sets of direction cosines, those of the reference and object waves (Fig. 4.6):

$$l_R = \cos \theta_R \qquad l_o = \cos \theta_o$$
$$m_R = \sin \theta_R \qquad m_o = \sin \theta_o \tag{4.29}$$

so that we can write

$$\delta(x, z) = k'z(l_R - l_o) + k'x(m_R - m_o). \tag{4.30}$$

Now, referring to Fig. 4.2, let the direction cosines of the fringes in the medium be given by l and m, where

$$l = \cos \alpha$$
$$m = \sin \alpha \tag{4.31}$$

so that

$$\tan \alpha = \frac{m}{l} = \frac{\cos \theta_o - \cos \theta_R}{\sin \theta_R - \sin \theta_o} = \frac{l_o - l_R}{m_R - m_o}. \tag{4.32}$$

Equation 4.30 may be written as

$$\delta(x, z) = \frac{k'}{l}(m_R - m_o)(lx - mz). \tag{4.33}$$

By using the relations (4.15) and (4.18) we can write

$$m_R - m_o = 2l \sin \frac{\varphi}{2} \tag{4.34}$$

so that

$$\delta(x, z) = 2k' \sin \frac{\varphi}{2} [lx - mz]. \tag{4.35}$$

Hence we have

$$|H(x, z)|^2 = a_o^2 + a_R^2 + 2a_o a_R \cos \omega_o (lx - mz), \tag{4.36}$$

where we have defined

$$\omega_o = 2k' \sin \frac{\varphi}{2}. \tag{4.37}$$

Thus we can write the pupil function as

$$G(x, z) = E_o[1 + M \cos \omega_o (lx - mz)]$$
$$= E_o\left\{1 + \frac{M}{2}[e^{i\omega_o(lx-mz)} + e^{-i\omega_o(lx-mz)}]\right\} \tag{4.38}$$

where E_o is the average exposure received by the recording medium and is proportional to $a_o^2 + a_R^2$. The quantity M is the exposure modulation and is proportional to $2a_o a_R/a_o^2 + a_R^2$.

4.2.3 The Bragg Condition

Substitution of Eq. 4.38 into 4.27 yields

$$U(l_i, m_i) = CE_o \int_{-H}^{H} \int_{-t/2}^{t/2} \left\{ e^{-ikx(m_i-m_c)}e^{-ikz(l_i-l_c)} + \frac{M}{2} \left[e^{i\omega_o(lx-mz)}e^{-ikx(m_i-m_c)} \right. \right.$$

$$\left. \left. \times\, e^{-ikz(l_i-l_c)} + e^{-i\omega_o(lx-mz)}e^{-ikx(m_i-m_c)}e^{-ikz(l_i-l_c)} \right] \right\} dx\, dz$$

which divides into three integrals,

$$U(l_i, m_i) = \frac{mCE_o}{2} \int_{-H}^{H} \int_{-t/2}^{t/2} e^{-ix(km_i-km_c-\omega_o l)}e^{-iz(kl_i-kl_c+\omega_o m)}\, dx\, dz$$

$$+ \frac{MCE_o}{2} \int_{-H}^{H} \int_{-t/2}^{t/2} e^{-ix(km_i-km_c+\omega_o l)}e^{-iz(kl_i-kl_c-\omega_o m)}\, dx\, dz$$

$$+ CE_o \int_{-H}^{H} \int_{-t/2}^{t/2} e^{-ikx(m_i-m_c)}e^{-ikz(l_i-l_c)}\, dx\, dz. \qquad (4.39)$$

Since

$$\int_{-H}^{H} e^{-iBx}\, dx = 2H\, \frac{\sin BH}{BH} = 2H\, \text{sinc } BH,$$

Equation 4.39 is readily integrated to

$$U(l_i, m_i) = \frac{MCE_o}{2} \left\{ \left[t\, \text{sinc}\left((kl_i - kl_c + \omega_o m)\frac{t}{2} \right) \right] \right.$$

$$\times \left[2H\, \text{sinc}\left((km_i - km_c - \omega_o l)H \right) \right] \right\}$$

$$+ \frac{MCE_o}{2} \left\{ \left[t\, \text{sinc}\left((kl_i - kl_c - \omega_o m)\frac{t}{2} \right) \right] \right.$$

$$\times \left[2H\, \text{sinc}\left((km_i - km_c + \omega_o l)H \right) \right] \right\}$$

$$+ 2CE_o Ht \left\{ \text{sinc}\left((kl_i - kl_c)\frac{t}{2} \right) \right\}$$

$$\times \left\{ \text{sinc}\left[(km_i - km_c)H \right] \right\}. \qquad (4.40)$$

This equation is the first-order solution to the problem of diffraction from a three-dimensional sinusoidal grating. The last term is a maximum when

$$l_i = l_c \quad \text{and} \quad m_i = m_c. \qquad (4.41)$$

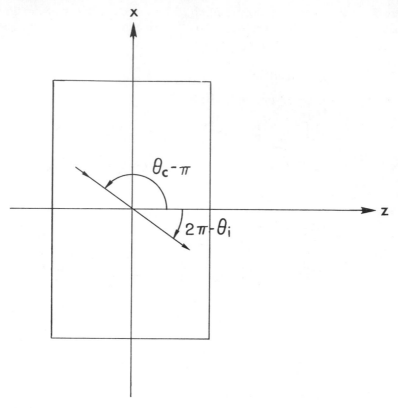

Fig. 4.7 The directly transmitted zero-order wave.

Since $l_c^2 + m_c^2 = 1$ and $l_i^2 + m_i^2 = 1$, the fact that $l_i = l_c$ implies $m_i = m_c$. This condition is illustrated in Fig. 4.7. Since we are using the Fraunhofer approximation, l_c and l_i, as defined by Eq. 4.23, now refer to the angles θ_c and θ_i shown in Fig. 4.8, that is,

$$\begin{array}{ll} l_c = \cos \theta_c & m_c = \sin \theta_c \\ & \text{and} \\ l_i = \cos \theta_i & m_i = \sin \theta_i. \end{array} \qquad (4.42)$$

This is just the directly transmitted or zero-order wave.

The first two terms of Eq. 4.40 are just the two first-order diffracted waves, corresponding to the primary and conjugate images. The first term is a maximum when the arguments of both sinc functions are zero, or when

$$l_i = l_c - \frac{\omega_o m}{k} \quad \text{and} \quad m_i = m_c + \frac{\omega_o l}{k}. \qquad (4.43)$$

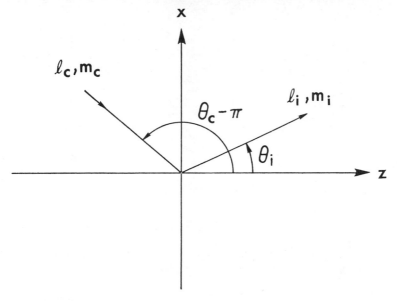

Fig. 4.8 Angles and direction cosines for the illuminating and diffracted waves.

We can write the grating equation as

$$d(\sin \theta_i - \sin \theta_c) = \lambda \qquad (4.44)$$

for the first diffracted order. Now since $m_c = \sin \theta_c$ and $m_i = \sin \theta_i$, the grating equation may be written

$$m_i - m_c = \frac{\lambda}{d} = \frac{2\pi\nu_s}{k} = \frac{\omega}{k} \qquad (4.45)$$

where ω is the radian spatial frequency of the grating in the x-direction. Now from Eqs. 4.18, 4.32, and 4.37 we have

$$\omega = \omega_o l \qquad (4.46)$$

and so the grating equation becomes

$$m_i - m_c = \frac{\omega_o l}{k}. \qquad (4.47)$$

Hence, we see that the second condition of Eq. 4.43 implies simply that there is a diffraction maximum in the direction predicted by the grating equation.

Referring to Fig. 4.9, we see that we can write the law of reflection in the form

$$l_c l + m_c m = l_i l + m_i m \qquad (4.48)$$

since $\cos i = l_c l + m_i m$ and $\cos r = l_i l + m_i m$.

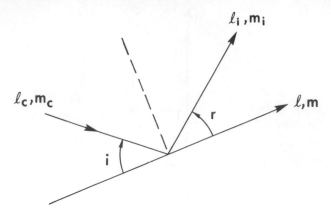

Fig. 4.9 How the reflection law ($r = i$) can be expressed in terms of the direction cosines.

Hence the law of reflection can be written as

$$l_c - l_i = (m_i - m_c)\frac{m}{l} \tag{4.49}$$

for an incident ray with direction cosines l_c, m_c being reflected from a plane with direction cosines l, m into the direction l_i, m_i. Since the first-order diffracted wave is in the direction required by Eq. 4.47, Eq. 4.49 may be written

$$l_c - l_i = \frac{\omega_o m}{k}, \tag{4.50}$$

which is just the first of the conditions in Eq. 4.43. Hence we see that there is a maximum for the first term of Eq. 4.40 in the direction l_i, m_i, such that the grating equation is satisfied simultaneously with an apparent reflection from the plane of the fringes lying in the l, m direction. These two conditions constitute the *Bragg condition* for diffraction from a three-dimensional grating. This is illustrated in Fig. 4.10. Similar arguments apply to the second term of Eq. 4.40. Note that it is impossible to have simultaneous maxima in both first orders. The first term of Eq. 4.40 is maximum when

$$\frac{-(l_i - l_c)}{m} = \frac{\omega_o}{k}$$

and

$$\frac{m_i - m_c}{l} = \frac{\omega_o}{k} \tag{4.51}$$

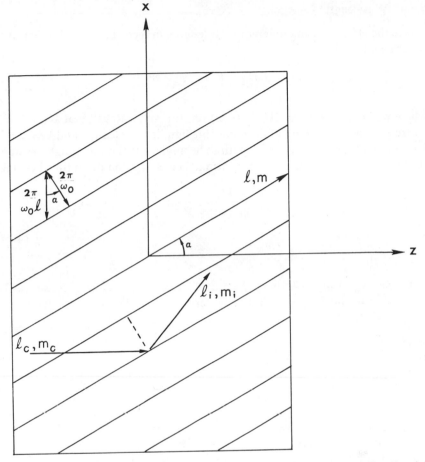

Fig. 4.10 The Bragg condition for diffraction. The entering wave in the direction (l_c, m_c) is diffracted in the direction (l_i, m_i) by the fringes lying in the direction (l, m).

and the second term is a maximum when

$$\frac{l_i - l_c}{m} = \frac{\omega_o}{k}$$

and

$$\frac{-(m_i - m_c)}{l} = \frac{\omega_o}{k}.$$

(4.52)

The directions of Eq. 4.52 are just the reverse of those in Eq. 4.51.

4.2.4 Wavelength Change

To see the effect of using different wavelengths in recording and illuminating, we write Eq. 4.51 as

$$\frac{-(l_i - l_c)}{m} = 2\frac{k'}{k}\sin\frac{\varphi}{2} = \frac{m_i - m_c}{l} \qquad (4.53)$$

where we have used $\omega_o = 2k'\sin(\varphi/2)$ (cf. Eq. 4.37). Recall that $k' = 2\pi/\lambda'$, where λ' is the wavelength of the recording light in the medium and $k = 2\pi/\lambda$; λ is here the illuminating wavelength. The angle φ is the total angle between the two beams used to record the grating. The Bragg condition thus becomes

$$\frac{l_c - l_i}{m} = 2\frac{\lambda}{\lambda'}\sin\frac{\varphi}{2}$$

and

$$\frac{m_i - m_c}{l} = 2\frac{\lambda}{\lambda'}\sin\frac{\varphi}{2} \qquad (4.54)$$

But since $m_R - m_o = 2l\sin(\varphi/2)$ (Eq. 4.34) and $l_R - l_0 = -2m\sin(\varphi/2)$, we can write the Bragg condition in terms of the direction cosines of the recording beams as

$$m_i - m_c = \frac{\lambda}{\lambda'}(m_R - m_o)$$

and

$$l_i - l_c = \frac{\lambda}{\lambda'}(l_R - l_o). \qquad (4.55)$$

From these equations we can see readily that for $\lambda = \lambda'$ and $l_c = -l_R$ (illuminating beam reversed in direction from the reference wave), $l_i = -l_o$; that is, the diffracted wave is traveling in a direction opposite to the object wave along the same path. This is the conjugate image term.

Using (4.52), we obtain a set of equations similar to (4.55) for the primary wave:

$$m_i - m_c = -\frac{\lambda}{\lambda'}(m_R - m_o)$$

$$l_i - l_c = -\frac{\lambda}{\lambda'}(l_R - l_o). \qquad (4.56)$$

Here we see that for $\lambda = \lambda'$ and $l_c = l_R$ (illuminating beam identical to the reference beam), $l_i = l_o$; that is, the diffracted wave is identical to the original object wave.

We may generalize to the case of a more complicated object by noting that the first term of Eq. 4.40 gives rise to a conjugate (real) image when all of the rays of the reference beam are reversed in direction and the same wavelength is used in recording and illuminating. Thus, if the reference wave is a diverging spherical wave, a real image will be formed in the object when the illuminating wave is spherical and converges to the same point in space from which the reference wave diverged.

The second term of Eq. 4.40 will yield a primary (virtual) image in the same position as the object if the illuminating wave is identical to the object wave.

There is further solution to Eqs. 4.55 and 4.56 for $l_c = l_o$ and $l_c = -l_o$, respectively. Either choice of illuminating wave then leads to $l_i = l_R$ or $l_i = -l_R$. This is a valid solution (for $\lambda = \lambda'$) for the case of a plane object wave as considered here. However, for a more complicated object wave, the illuminating wave will not, in general, match exactly the object wave. The condition $l_c = l_o$ is then only satisfied for a single component of the plane wave spectrum of the object wave, and no complete image will be formed.

If the wavelength of the illuminating wave differs from that of the recording waves ($\lambda \neq \lambda'$), it is still possible to find solutions to Eqs. 4.55 and 4.56. However, when we make the generalization from a single plane object wave to a more general object wave represented by a spectrum of plane waves, then we see that the Bragg condition cannot be satisfied simultaneously by all components. We can see this by differentiating Eq. 4.55. For example,

$$\Delta l_c = \frac{\Delta \lambda}{\lambda'} (l_R - l_o). \tag{4.57}$$

Here we see that the change in direction Δl_c of the illuminating beam required by a change in wavelength $\Delta \lambda$ is a function of the direction of the object wave l_o. Hence, no single illuminating wave can simultaneously satisfy the Bragg condition for a whole spectrum of object waves. In general, no complete image will be formed with a change in wavelength.

A further restriction on the allowable change in wavelength results when we require that no light be diffracted at angles greater than 90°. Hence, from Eq. 4.45 we must have the grating spacing $d \geq \lambda$. Writing d as $2\pi/\omega$ with ω given by Eq. 4.18, we arrive at the requirement

$$\frac{\lambda}{\lambda'} \leq \frac{1}{m_R - m_o}. \tag{4.58}$$

4.2.5 Orientation Sensitivity

All of the foregoing conclusions apply only in the case of truly thick holograms where the fringe spacing is very much smaller than the thickness of the recording medium. For most holograms made on photographic emulsions, this is only approximately true, and there is generally a finite range of directions over which an object wave can be reconstructed. To determine just how large this range is, we return to Eq. 4.40. If we consider only the second term, we see that the amplitude diffracted into the first order is

$$U(l_i, m_i) = MCE_o Ht \left\{ \text{sinc} \left[(kl_i - kl_c - \omega_o m) \frac{t}{2} \right] \right\}$$
$$\times \{ \text{sinc} \left[(km_i - km_c + \omega_o l)H \right] \} \quad (4.59)$$

where t is the thickness of the hologram and H its dimension. We now construct a ratio I/I_o where

$$I_o = [MCE_o Ht]^2 \quad (4.60)$$

is the irradiance of the diffracted wave when the Bragg condition is satisfied, and

$$I = I_o \text{sinc}^2 \left[(kl_i - kl_c - \omega_o m) \frac{t}{2} \right] \quad (4.61)$$

is the irradiance of the diffracted wave in a direction determined by the grating equation, but the incident and diffracted rays do not appear to be reflected from the Bragg planes. We have then

$$\frac{I}{I_o} = \text{sinc}^2 \left[(kl_i - kl_c - \omega_o m) \frac{t}{2} \right]$$
$$= \text{sinc}^2 \left\{ [k(l_i - l_c) + k'(l_R - l_o)] \frac{t}{2} \right\}. \quad (4.62)$$

When the argument of the sinc function in Eq. 4.62 equals π, the relative irradiance in the diffracted beam is zero. This indicates, for example, the minimum angular rotation of the hologram that will extinguish the image. Alternatively, it indicates the minimum angular displacement of the reference beam in order to record more than one hologram that can be read out separately without overlapping images.

Let us first assume that $\lambda = \lambda'$, so that the argument of the sinc function because

$$(l_i - l_c + l_R - l_o) \frac{\pi t}{\lambda}. \quad (4.63)$$

Now a small rotation of the hologram results in a change Δl_c in l_c with the corresponding change in direction of the diffracted wave Δl_i in l_i. The argument of the sinc function now becomes

$$(l_i + \Delta l_i - l_c - \Delta l_c + l_R - l_o) \frac{\pi t}{\lambda} \quad (4.64)$$

and when this equals π, I/I_o becomes zero:

$$(l_i + \Delta l_i - l_c - \Delta l_c + l_R - l_o)\frac{\pi t}{\lambda} = \pi. \tag{4.65}$$

If we assume that we started with the illuminating wave identical to the reference wave ($l_c = l_R$), and therefore the diffracted wave identical to the object wave ($l_i = l_o$), Eq. 4.65 becomes

$$\Delta l_i - \Delta l_c = \frac{\lambda}{t}. \tag{4.66}$$

Similarly, since we are assuming that the grating equation is satisfied, the argument of the second sinc function of Eq. 4.59 remains zero and therefore we must have

$$\Delta m_i - \Delta m_c = 0. \tag{4.67}$$

Now since

$$\begin{aligned}
\Delta l_i &= -\sin\theta_i\,\Delta\theta_i = -m_i\,\Delta\theta_i \\
\Delta l_c &= -\sin\theta_c\,\Delta\theta_c = -m_c\,\Delta\theta_c \\
\Delta m_i &= \cos\theta_i\,\Delta\theta_i = l_i\,\Delta\theta_i \\
\Delta m_c &= \cos\theta_c\,\Delta\theta_c = l_c\,\Delta\theta_c
\end{aligned} \tag{4.68}$$

Eq. 4.67 gives

$$\Delta\theta_i = \frac{l_c}{l_i}\Delta\theta_c. \tag{4.69}$$

Substitution into Eq. 4.66 then yields

$$\Delta\theta_c = \frac{\lambda}{t}\left(\frac{1}{m_c - m_i l_c / l_i}\right) \tag{4.70}$$

and since $(l_i, m_i) = (l_o, m_o)$ and $(l_c, m_c) = (l_R, m_R)$, we have

$$\Delta\theta_c = \frac{\lambda}{t}\left[\frac{l_o}{m_R l_o - m_o l_R}\right] \tag{4.71}$$

with a similar expression for the image shift,

$$\Delta\theta_i = \frac{\lambda}{t}\left[\frac{l_R}{m_R l_o - m_o l_R}\right]. \tag{4.72}$$

Equation 4.71 gives the angular rotation in radians required to extinguish the image formed with a hologram of thickness t and made and illuminated with light of wavelength λ *in the medium*. The hologram was recorded with two plane waves with direction cosines (l_o, m_o) and (l_R, m_R). For the case where

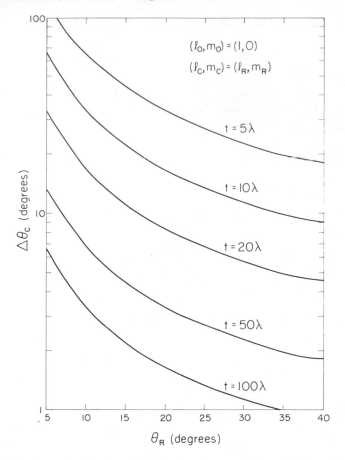

$(l_o, m_o) = (1, 0)$

$(l_c, m_c) = (l_R, m_R)$

$t = 5\lambda$

$t = 10\lambda$

$t = 20\lambda$

$t = 50\lambda$

$t = 100\lambda$

Fig. 4.11 The change in direction of the illuminating wave $\Delta\theta_c$ required to extinguish the diffracted wave as a function of the angle of incidence of the reference wave, θ_R. The curves are plotted for $\theta_o = 0$ and for several values of hologram thickness. The wavelength and angles refer to values in the medium.

$(l_o, m_o) = (1, 0)$ (object wave incident normally), Eq. 4.71 becomes

$$\Delta\theta_c = \frac{\lambda}{t}\frac{1}{m_R} = \frac{\lambda}{t}\csc\theta_R. \tag{4.73}$$

Figure 4.11 shows a plot of this function for several values of λ/t. The values of both λ and θ_R refer to those in the recording medium. Not all angles θ_R are readily accessible, because an angle of incidence of 90° outside the recording medium results in an angle $\theta_R \simeq 40°$ inside for an index of 1.5. Angles greater than 90° correspond to the reference wave entering from the back side of the hologram. Figure 4.12 indicates the conditions assumed for calculating

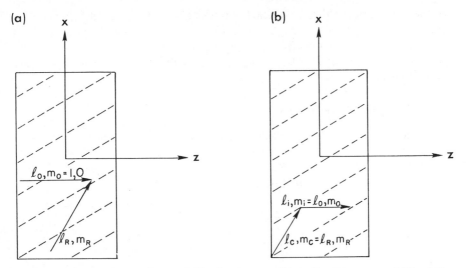

Fig. 4.12 Assumed recording and illuminating conditions for the curves of Fig. 4.11. (*a*) Recording. (*b*) Illuminating.

the curves of Fig. 4.11. In 4.12*a*, the hologram is recorded with the (plane) object wave incident normally so that $\theta_o = 0$ and $(l_o, m_o) = (1, 0)$. The reference wave makes an angle θ_R with the *z*-axis. Fringes are recorded lying in the direction (l, m), which is the bisector of the angle between the object and reference waves. The radian spatial frequency ω of the recorded grating depends on the angle between these two waves:

$$\omega = \frac{2\pi}{\lambda}(\sin \theta_R - \sin \theta_o)$$

$$= \frac{2\pi}{\lambda}\sin \theta_R \quad \text{for} \quad \lambda = \lambda', \; \theta_o = 0.$$

Hence small values of θ_R imply low spatial frequencies. Figure 4.12*b* shows that we are also assuming that the grating is illuminated with a plane wave in the same direction as the reference wave: $(l_c, m_c) = (l_R, m_R)$. The diffracted wave is then in the same direction as the object wave: $(l_i, m_i) = (l_o, m_o)$.

Figure 4.11 shows that for small θ_R, a relatively large change in θ_c is required to extinguish the diffracted wave. Hence the irradiance of the diffracted wave is relatively insensitive to the orientation of the hologram (θ_c can be changed by rotating the hologram) if the reference wave makes only a small angle with the object wave. The curves of Fig. 4.11 are symmetrical about $\theta_R = 90°$, so the irradiance of the diffracted wave is also relatively insensitive to orientation of the reference wave is incident from the back side of the hologram in a direction nearly opposite that of the object beam (reflection or Lippmann holograms, Section 4.3). On the other hand, the amount of flux diffracted is very sensitive to hologram orientation when the object and reference waves meet at a large angle, so that the recorded fringes are very closely spaced.

4.2.6 Wavelength Sensitivity

Next, let us consider the effect of illuminating the hologram with light of a wavelength different from that used for recording. From Eq. 4.59 we see that the diffracted amplitude corresponding to the primary wave is proportional to the product of two sinc functions of arguments

$$(kl_i - kl_c - \omega_o m)\frac{t}{2} \tag{4.74}$$

and

$$(km_i - km_c + \omega_o l)H. \tag{4.75}$$

From Eq. 4.37 we have

$$\omega_o = 2k' \sin \frac{\varphi}{2} \tag{4.76}$$

and from (4.35) we have

$$2l \sin \frac{\varphi}{2} = m_R - m_o$$

$$-2m \sin \frac{\varphi}{2} = l_R - l_o \tag{4.77}$$

so that Eqs. 4.74 and 4.75 may be written as

$$[k(l_i - l_c) + k'(l_R - l_o)]\frac{t}{2} \tag{4.78}$$

$$[k(m_i - m_c) + k'(m_R - m_o)]H. \tag{4.79}$$

Now assume that we start off with $\lambda = \lambda'$ and the condition for a maximum for which $(l_i, m_i) = (l_o, m_o)$ and $(l_c, m_c) = (l_R, m_R)$. We now ask what change in wavelength, with the illuminating wave in the same direction as the reference wave $(l_c = l_R)$, will cause Eq. 4.62 to vanish? This will occur when the argument of the sinc function (Eq. 4.78) equals π. A change in wavelength results in a change in direction of the diffracted wave, so that as $k' \to k + \Delta k$, $l_i \to l_i + \Delta l_i$. We then have as the condition for a zero

$$[k(l_i + \Delta l_i - l_c) + k(l_R - l_o) + \Delta k(l_R - l_o)]\frac{t}{2} \tag{4.80}$$

and

$$[k(m_i + \Delta m_i - m_o) + k(m_R - m_o) + \Delta k(m_R - m_o)]H. \qquad (4.81)$$

But since we have chosen the condition $(l_i, m_i) = (l_o, m_o)$ and $(l_c, m_c) = (l_R, m_R)$, these reduce to

$$\Delta k(l_R - l_o) + k \Delta l_i = \frac{2\pi}{t} \qquad (4.82)$$

and

$$\Delta k(m_R - m_o) + k \Delta m_i = 0 \qquad (4.83)$$

where $k = 2\pi/\lambda$ and $\Delta k = -(2\pi/\lambda^2) \Delta\lambda$. Using Eq. 4.68 we obtain the change in wavelength required for extinction of the image:

$$\Delta\lambda = \frac{\lambda^2}{t}\left[\frac{l_o}{1 - l_o l_R - m_o m_R}\right]. \qquad (4.84)$$

This change in wavelength corresponds to a change in the direction of the diffracted beam given by

$$\Delta\theta_i = \frac{\lambda}{t}\left[\frac{m_R - m_o}{1 - l_o l_R - m_o m_R}\right]. \qquad (4.85)$$

Figure 4.13 shows curves of $\Delta\lambda$ for several values of λ/t (in the medium) for the case $(l_o, m_o) = (1, 0)$.

On the basis of Figs. 4.11 and 4.13, we can, following Leith et al. [2], classify holograms into three general categories, depending on the value of φ, the total angle between reference and object waves.

Class A: φ is small, so that ω_o is small compared to $1/t$. In this case, the diffracted irradiance is relatively insensitive to change in illuminating wavelength or direction and the recording medium behaves as a two-dimensional medium.

Class B: φ is moderately large (10–120° for photographic emulsions). In this region the diffracted irradiance is most sensitive to change in the direction of the illuminating wave, so that hologram alignment is critical. A number of different holograms may be stored on a single photographic plate, each of which may be read out without interference from the others, if each is made with the reference beam in a different direction. The diffracted irradiance is fairly sensitive to changes in illuminating wavelength, and one can suppress wavelengths separated by several hundred angstroms from the recording wavelength with most photographic emulsions.

Class C: φ is large, with object and reference waves entering from opposite sides of the recording medium. In this case the recorded fringes (Bragg planes) lie nearly parallel to the x-y plane of the hologram. This type of hologram is

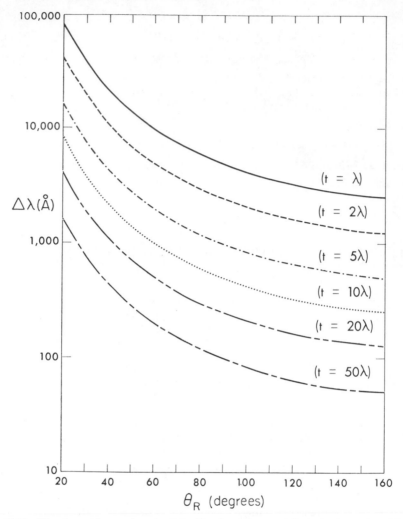

Fig. 4.13   The change in wavelength of the illuminating wave $\Delta\lambda$ required to extinguish the diffracted wave as a function of the angle of incidence of the reference wave, θ_R. The curves are plotted for $\theta_o = 0$ and for several values of hologram thickness. The wavelength and angles refer to values in the medium.

most sensitive to a change in wavelength, the fringes being separated by only about one-half the wavelength in the medium. Holograms of this class may be illuminated with white light, since it will act like an interference filter, producing a diffracted wave for only a narrow band of wavelengths.

This type of hologram is, however, relatively insensitive to a change in direction of the illuminating wave. A rotation of the hologram does not cause the extinction of the image, but rather a change in color so as to maintain the Bragg condition (Eqs. 4.55 and 4.56). This class of hologram was first described by Denisyuk [4], who noted its similarity to Lippmann color photography; it is sometimes called a Lippmann or reflection hologram. Because of the importance of this type of hologram we shall describe it in detail in the next section.

4.3   REFLECTION HOLOGRAMS [4]

Reflection holograms are made by allowing the reference and object beams to enter the recording medium from opposite sides, so that they are traveling in approximately opposite directions (Fig. 4.14). The interference of these two beams forms a stationary standing wave pattern in the recording medium. After processing, an image exists in the recording medium with a density that is proportional to the irradiance of the standing wave pattern. The fringes formed are approximately planes perpendicular to the z-axis. If such a hologram is illuminated with a beam of light similar to the reference beam, then it will reflect a portion of this light that is identical to the original object wave. The reconstruction is viewed in reflection rather than in transmission as with the usual hologram. White light may also be used for reconstruction, since the wavelength sensitivity of this type of hologram is very high (Fig. 4.13).

To consider the theory of the process, we write the object and reference waves as $O_o(x, y, z)e^{i\varphi_o(x,y,z)}$ and $R_o(x, y, z)e^{i\varphi_R(x,y,z)}$, respectively. The total field in the volume of the recording medium is

$$H(x, y, z) = O_o e^{i\varphi_o} + R_o e^{i\varphi_R}$$

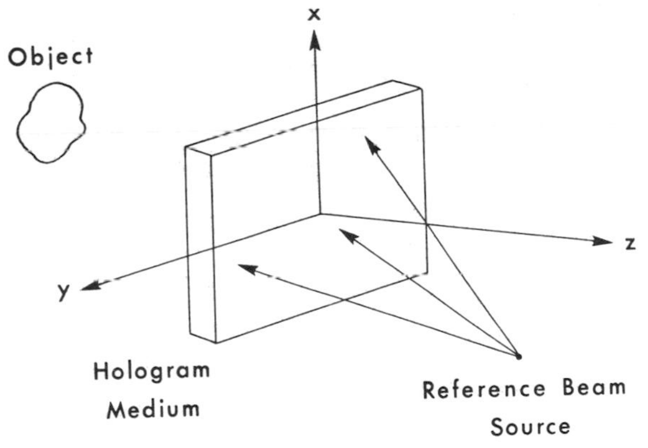

Fig. 4.14   Illustrating the general arrangement for recording a reflection hologram.

and the resultant irradiance of the standing wave pattern is

$$|H|^2 = O_o^2 + R_o^2 + 2O_oR_o \cos{(\varphi_o - \varphi_R)}. \qquad (4.86)$$

The fringes are the loci of constant phase difference

$$\delta = \varphi_o - \varphi_R \qquad (4.87)$$

so that we can write for a fringe

$$|H|^2 = O_o + R_o^2 + 2O_oR_o \cos{\delta}. \qquad (4.88)$$

Now assume that the recording medium is such that the resultant recorded signal D is proportional to $|H|^2$:

$$D = \chi |H|^2. \qquad (4.89)$$

Assuming that the dielectric constant ϵ of the medium is related to D, we can, for small D, expand $\epsilon(D)$ into a series

$$\epsilon(D) = \epsilon|_{D=0} + D\frac{\partial \epsilon}{\partial D}\bigg|_{D=0} + \cdots$$

$$= \epsilon_o + \gamma D \qquad (4.90)$$

where $\epsilon|_{D=0}$ is the dielectric constant of the unexposed recording medium. Substituting Eq. 4.88 into 4.89, and then substituting this value of D into Eq. 4.90, we obtain for the dielectric constant

$$\epsilon = \epsilon_o + \gamma\chi[O_o^2 + R_o^2 + 2O_oR_o \cos{\delta}]. \qquad (4.91)$$

Let us separate the infinitely thin layer enclosed between the isophase surface described by $\delta = $ const, and the isophase surface described by $\delta + d\delta = $ const. Such a layer may be regarded as the interface between two media, where $d\epsilon$ equals the difference between the dielectric constants of these media. This interface will then behave as a mirror surface. Writing the amplitude reflection coefficient as

$$dr = \frac{\sqrt{(\epsilon + d\epsilon)/\epsilon} - 1}{\sqrt{(\epsilon + d\epsilon)/\epsilon} + 1} \qquad (4.92)$$

and using the binomial expansion for the square roots, we find

$$dr \simeq \frac{d\epsilon}{4\epsilon}, \qquad (4.93)$$

with $d\epsilon = 2\gamma\chi O_oR_o \sin{\delta}\, d\delta$. Hence we can write for the amplitude reflectance of this layer

$$dr = CO_o \sin{\delta}\, d\delta \qquad (4.94)$$

where

$$C \equiv \frac{\gamma \chi R_o}{2\epsilon} \qquad (4.95)$$

Thus, the amplitude reflected from this layer is proportional to the amplitude of the light scattered by the object. Let us now see how a beam of light interacts with this layer. Let a beam of light identical to the reference beam $(R_o e^{i\varphi_R})$ be incident on the hologram. Following Denisyuk, we take advantage of an assumed low density of the image in the recording medium and determine the form of the radiation reflected from each individual isophase layer, in order later to combine these to determine the form of the radiation reflected by the entire hologram. Multiplying the incident wave $R_o e^{i\varphi_R}$ by the amplitude reflectance dr, we find the amplitude of the wave at the surface of the layer:

$$d\psi = CR_o e^{i\varphi_R} O_o \sin \delta \, d\delta. \qquad (4.96)$$

But from Eq. 4.87, $\varphi_R = \varphi_o - \delta$ on the isophase surface, so that

$$d\psi = (CR_o \sin \delta \, d\delta \, e^{-i\delta}) O_o e^{i\varphi_o}. \qquad (4.97)$$

We see that the wave reflected from the isophase layer is identical to the original object wave except for the constant multiplier in parentheses in Eq. 4.97. An observer viewing this wave will see a three-dimensional virtual image of the object. A real image is formed when the illuminating wave converges to a real image of the reference beam source.

By summing the waves reflected by all of the elementary isophase layers, it is possible to show that this type of hologram also reproduces the spectral composition of the object wave. Thus the object wave can be reconstructed even if the hologram is illuminated with white light. To show this, the illuminating wave is written as

$$\int_0^\infty R(k') e^{ik'\mathscr{L}(\mathbf{r})} \, dk' \qquad (4.98)$$

where $\mathscr{L}(\mathbf{r})$ describes the optical path of the light and satisfies the eikonel equation

$$[\nabla \mathscr{L}(\mathbf{r})]^2 = n^2(\mathbf{r}) \qquad (4.99)$$

and $\mathbf{r}$ is the position vector. Substituting Eq. 4.98 into 4.96 in place of $R_o e^{i\varphi_R}$ we have

$$d\psi = CO_o \sin \delta \, d\delta \int_0^\infty R(k') e^{ik'\mathscr{L}(\mathbf{r})} \, dk'. \qquad (4.100)$$

But

$$\delta = \varphi_o - \varphi_R = k\mathscr{L}_o(\mathbf{r}) - k\mathscr{L}(\mathbf{r}) = k\rho \qquad (4.101)$$

where $\mathscr{L}_o(\mathbf{r})$ describes the optical path of the object beam and $\mathscr{L}(\mathbf{r})$ that of the reference beam. We are assuming that the illuminating beam (Eq. 4.98) follows the same path as the reference beam. The quantity ρ is the optical path difference between the two beams, and k is of course $2\pi/\lambda$. Using Eq. 4.101 we can write

$$d\psi = kCO_o \sin \rho \, d\rho \int_0^\infty R(k')e^{ik'\mathscr{L}_o(\mathbf{r})}e^{-ik'\rho} \, dk'. \tag{4.102}$$

Summing the contributions from each isophase layer between the layers described by ρ_1 and ρ_2 (assuming weak reflection), the reflected wave of interest becomes

$$\psi = kCO_o \int_{\rho_1}^{\rho_2} \frac{e^{ik\rho} - e^{-ik\rho}}{2i} \, d\rho \int_0^\infty R(k')e^{ik'\mathscr{L}_o(\mathbf{r})}e^{-ik'\rho} \, dk'. \tag{4.103}$$

Changing the order of integration, this becomes

$$\psi = \frac{-ikCO_o}{2} \int_0^\infty R(k')e^{ik'\mathscr{L}_o(\mathbf{r})} \, dk' \int_{\rho_1}^{\rho_2} [e^{i\rho(k-k')} - e^{-i\rho(k+k')}] \, d\rho. \tag{4.104}$$

Carrying out the ρ-integration we obtain

$$\psi = \frac{-ikCO_o}{2} \int_0^\infty R(k')e^{ik'\mathscr{L}_o(\mathbf{r})}[\delta_1(k + k') - \delta_1(k - k')] \, dk' \tag{4.105}$$

where

$$\delta_1(k - k') = \frac{e^{i\rho_2(k-k')} - e^{i\rho_1(k-k')}}{i(k - k')} \tag{4.106}$$

which can be written

$$\delta_1(k - k') = \frac{2e^{(i/2)(\rho_1+\rho_2)(k-k')}}{k - k'} \sin\left[\frac{(\rho_2 - \rho_1)(k - k')}{2}\right]. \tag{4.107}$$

The functions $\delta_1(k + k')$ and $\delta_1(k - k')$ have the properties of Dirac delta functions. The function $\delta_1(k - k')$ has a maximum for $k = k'$, where it assumes the value $\rho_2 - \rho_1$. Similarly, $\delta_1(k + k')$ has its maximum for $k = -k'$. The widths of these functions are given by

$$\Delta k = \frac{2\pi}{\rho_2 - \rho_1}. \tag{4.108}$$

The thickness of the hologram is assumed much larger than a wavelength; therefore $\rho_2 - \rho_1 \gg 1/k$, and the delta functions may be replaced by rectangular pulses of width Δk and height $\rho_2 - \rho_1$. Making this substitution, we find

$$\psi(\mathbf{r}) = \frac{-ikCO_o(\mathbf{r})}{2} [\rho_2 - \rho_1] \int_{k-\Delta k/2}^{k+\Delta k/2} R(k')e^{ik'\mathscr{L}_o(\mathbf{r})} \, dk'. \tag{4.109}$$

Therefore we see that the reflected wave is composed of wavelengths that differ little from the wavelength which exposed the hologram: its phase agrees with the phase of the wave scattered by the object; its amplitude is proportional to the amplitude of the object wave and the optical path $\rho_2 - \rho_1$ which the reconstructed wave travels through the hologram.

Making the appropriate changes of sign, one may also show that a conjugate image of the object is formed. The zero-order image is the light that is specularly reflected from the boundary of the hologram medium.

We thus see that a hologram made in this manner has some very interesting properties, the most noteworthy being that white-light reconstruction is possible. This was expected because of the great wavelength sensitivity shown in Fig. 4.13. Pennington and Lin [5] have used this effect to produce a color hologram. This was done by using a reference beam composed of red and blue light. A color transparency was illuminated by the same combination. When the resulting hologram was illuminated with white light, a good two-color reproduction of the transparency was obtained. The method has also been extended to three colors, and color holograms represent a large portion of present-day research in holography.

REFERENCES

[1] C. B. Burckhardt, *J. Opt. Soc. Am.*, **56**, 1502 (1966).

[2] E. N. Leith, A. Kozma, J. Upatnieks, J. Marks, and N. Massey, *Appl. Opt.*, **5**, 1303 (1966).

[3] E. G. Ramberg, *R.C.A. Rev.*, **27**, 467 (1966).

[4] Y. N. Denisyuk, *Optics and Spectroscopy*, **15**, 279 (1963).

[5] K. S. Pennington and L. H. Lin, *Appl. Phys. Letters*, **7**, 56 (1965).

5 Phase Holograms

5.0 INTRODUCTION

Thus far we have considered only absorption holograms—those for which the light of the illuminating beam is absorbed in correspondence with the recording exposure. In other words, we have assumed that the transmission function $t(x)$ was a real valued function. We now wish to show that holograms can also be formed on recording media for which $t(x)$ is a complex function, that is, the hologram alters the *phase* of the illuminating wave in correspondence with the recording exposure. Such a hologram is called a "phase hologram" and has some interesting and important properties.

In order to demonstrate most clearly the principles of phase holograms, we consider only the case of pure phase modulation in which $t(x)$ is pure imaginary. Holograms can, of course, be produced with $t(x)$ having both real and imaginary parts—some absorption and some phase modulation. Indeed, most of the holograms produced are probably of this type since it is difficult to eliminate phase modulation completely.

Pure phase holograms can be made by many methods. The earliest reference to phase holograms describes their preparation by contact printing to Gabor holograms, employing the Carbro process to produce relief images in transparent gelatin on glass substrates [1]. Phase holograms may be produced by contact printing (or by direct recording with blue or green laser light) onto a resist material. They may be produced by using a thermoplastic material as the recording medium [2]. It is also possible to produce phase holograms with conventional silver halide photographic emulsions, either by utilizing the relief image or index change, or both. Relief images are an emulsion thickness variation corresponding to the image, caused by the differential tanning action of the developer [3]. These thickness variations result in the desired spatial phase modulation of the transmitted light. To obtain a pure phase modulation using this technique, one can bleach out the exposed silver grains so that no density variation remains [4], or one can metallize the emulsion surface and view the hologram image in reflection [5]. Phase holograms may be produced by utilizing a change in the index of refraction of the recording medium. In the photoresist materials, a change in index results from a photopolymerization of the resist material upon exposure. There are now no thickness variations corresponding to exposure, but the signal is recorded in the form of index variations that impose the desired phase modulation onto the illuminating beam. A bleached photographic emulsion is most commonly a hologram of this type.

These are just some of the methods for producing phase holograms that are known at this time; there will undoubtedly be more methods to come. The main interest in phase holograms stems from their very high *diffraction efficiency*, the ratio of incident to usable diffracted flux. Burckhardt has shown that the efficiency of a thick dielectric grating can be as high as 100% [6]. For a thin phase grating, the efficiency can be as high as 34% [7]. These numbers are to be compared with a little over 6% for the usual absorption type of gratings (or hologram). A more complete discussion of diffraction efficiency of the various hologram types is presented in Chapter 7.

5.1 ANALYSIS

As stated above, we will, for simplicity, restrict our attention to pure phase holograms, that is, ones for which only the phase and not the amplitude of the illuminating wave is affected.

To begin, we suppose that the recording medium is such that an exposure $E(x)$ results in either a change in index $n(x)$ of the medium or a change in thickness $h(x)$ of the medium. If the exposed hologram is illuminated with a wave $C(x)$, then the transmitted wave will have a phase variation imposed on it of the form $knh(x)$ or $khn(x)$ for the relief or index types, respectively. This imposed phase modulation yields various diffracted orders, the positive or negative first orders producing the desired reconstructions. For thick holograms, one can arrange things so as to produce only one diffracted order, or one can eliminate the zero order, by a suitable adjustment of the phase shift.

Next we expose the recording medium to an object wave described by

$$O(x) = O_o e^{i\varphi_o(x)} \tag{5.1}$$

and a plane reference wave

$$R(x) = R_o e^{i\beta x} \tag{5.2}$$

where $\beta = k \sin \alpha_R$, as shown in Fig. 5.1. Let us further assume that the final phase modulation to be imposed on the illuminating wave $C(x)$ is simply

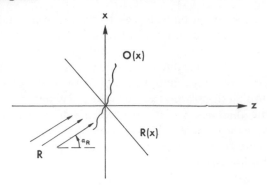

Fig. 5.1   Recording an off-axis phase hologram.

proportional to the exposure $E(x)$. Call this phase modulation $\varphi(x)$ so that we can write

$$\varphi(x) = \gamma E(x) \tag{5.3}$$

where γ is the proportionality constant and $E(x)$ can be written as

$$
\begin{aligned}
E(x) &= |O(x) + R(x)|^2 \\
&= O_o^2 + R_o^2 + O_o R_o e^{-i(\varphi_o - \beta x)} + O_o R_o e^{i(\varphi_o - \beta x)} \\
&= O_o^2 + R_o^2 + 2O_o R_o \cos{(\varphi_o - \beta x)}.
\end{aligned}
\tag{5.4}
$$

The complex transmittance function then becomes

$$
\begin{aligned}
T_a(x) &= T_o e^{i\varphi(x)} = T_o e^{i\gamma E(x)} \\
&= T_o \exp{(i\gamma O_o^2)} \exp{(i\gamma R_o^2)} \exp{\{2i\gamma O_o R_o[\cos{(\varphi_o - \beta x)}]\}}.
\end{aligned}
\tag{5.5}
$$

For simplification of notation, we write

$$
\begin{aligned}
K &\equiv T_o \exp{(i\gamma O_o^2)} \\
a &\equiv 2\gamma O_o R_o \\
\theta &\equiv \varphi_o - \beta x
\end{aligned}
\tag{5.6}
$$

so that

$$T_a(x) = K e^{ia \cos \theta}. \tag{5.7}$$

Using the Bessel function expansions

$$\cos{(a \cos \theta)} = J_o(a) + 2 \sum_{n=1}^{\infty} (-1)^n J_{2n}(a) \cos{2n\theta}$$

$$\sin{(a \cos \theta)} = 2 \sum_{n=0}^{\infty} (-1)^{n+2} J_{2n+1}(a) \cos{[(2n + 1)\theta]}$$

$$\tag{5.8}$$

and the relation

$$e^{ix} = \cos x + i \sin x \tag{5.9}$$

we can write

$$T_a(x) = K\left\{J_o(a) + 2\sum_{n=1}^{\infty}(-1)^n J_{2n}(a)\cos 2n\theta \right.$$

$$\left. + 2i\sum_{n=0}^{\infty}(-1)^{n+2}J_{2n+1}(a)\cos[(2n+1)\theta]\right\}. \quad (5.10)$$

The $J_n(a)$ are Bessel functions of the first kind. Each $J_n(a)$ is the amplitude of the nth diffracted order. Note that in general all orders are present, unlike the case of an absorption hologram, where the sinusoidal amplitude modulation led to only the positive or negative first orders. The term of (5.10) leading to the images of interest is

$$K[2iJ_1(a)\cos\theta], \quad (5.11)$$

which can be written as

$$2iKJ_1(a)\left[\frac{e^{i\theta}+e^{-i\theta}}{2}\right] = iKJ_1(a)[e^{i(\varphi_o-\beta x)}+e^{-i(\varphi_o-\beta x)}]. \quad (5.12)$$

These two terms represent the primary and conjugate image waves. The transmission term leading to the primary image is

$$T_p(x) = T_o\exp[i(\gamma O_o^2 + \gamma R_o^2 + \pi/2)]J_1(2\gamma O_o R_o)\exp[i(\varphi_o - \beta x)]. \quad (5.13)$$

If the hologram is illuminated with a wave $C(x) = R(x)$ (Fig. 5.2), we obtain the primary image wave:

$$\psi_p(x) = C(x)T_p(x) = R_o e^{i\beta x}T_p(x)$$

$$= T_o\exp[i(\gamma O_o^2 + \gamma R_o^2 + \pi/2)]R_o J_1(2\gamma O_o R_c)e^{i\varphi_o}. \quad (5.14)$$

Now since

$$J_1(x) = \tfrac{1}{2}x - \frac{(\tfrac{1}{2}x)^3}{1^2\cdot 2} + \frac{(\tfrac{1}{2}x)^5}{1^2\cdot 2^2\cdot 3} - \cdots$$

$$\approx \tfrac{1}{2}x \quad \text{for small } x, \quad (5.15)$$

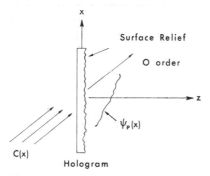

Fig. 5.2 Diffraction from a phase hologram.

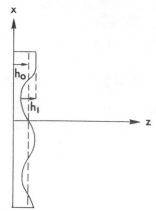

Fig. 5.3   Relief image parameters.

we have, if $2\gamma O_o R_o$ is small,

$$\psi_p(x) \cong \frac{T_o \gamma}{2} \exp\left[i(\gamma O_o^2 + \gamma R_o^2 + \pi/2)\right] R_o^2 O_o e^{i\varphi_o}, \tag{5.16}$$

which is seen to be, aside from the unimportant phase and amplitude factors, just the original object wave. If the product $2\gamma O_o R_o$ is large, some amplitude distortion will be present. From Eqs. 5.3 and 5.4, we see that $2\gamma O_o R_o$ is the amplitude of the phase modulation:

$$\varphi(x) = \gamma O_o^2 + \gamma R_o^2 + 2\gamma R_o O_o \cos(\varphi_o - \beta x). \tag{5.17}$$

Hence if the phase hologram is of the relief image type such as produced with a thermoplastic material, we might have, for example,

$$\varphi(x) = \gamma k n h(x) = \gamma O_o^2 + \gamma R_o^2 + 2\gamma O_o R_o \cos(\varphi_o - \beta x), \tag{5.18}$$

which can be written in the form

$$\varphi(x) = \gamma k n \left[h_o + \frac{h_1}{2} \cos(\varphi_o - \beta x) \right] \tag{5.19}$$

where h_1 is the peak-to-peak variation of the hologram thickness and the average thickness is h_0, as shown in Fig. 5.3. A distortion-free primary (or conjugate) image may be formed with a phase hologram if the phase modulation is small; in the above example, h_1 should be kept small.

5.2   DIFFRACTION EFFICIENCY AND NOISE

As mentioned in the Introduction, the diffraction efficiency of phase holograms, either thick or thin, is inherently higher than that of amplitude holograms. Unfortunately, it is also true that phase holograms are inherently more noisy than amplitude holograms. The reason for this can be explained as follows. The exposure at the hologram plane due to a reference beam R and an object beam O is given by

$$E = |O|^2 + |R|^2 + RO^* + R^*O. \qquad (5.20)$$

Again assuming that the final phase modulation imposed onto the illuminating wave is simply proportional to the exposure as in Eq. 5.3, the complex transmittance of the phase hologram becomes, as in Eq. 5.5,

$$T_a(x) = T_o e^{i\varphi(x)} = T_o e^{i\gamma |O|^2} e^{i\gamma |R|^2} e^{i\gamma RO^*} e^{i\gamma R^*O}. \qquad (5.21)$$

Expanding each of the exponentials into series form gives

$$T_a(x) = T_o[1 + i\gamma |O|^2 + \cdots][1 + i\gamma |R|^2 + \cdots]$$
$$\times [1 + i\gamma RO^* + \cdots][1 + i\gamma R^*O + \cdots]. \qquad (5.22)$$

If we now assume that R is given by Eq. 5.2 and that R_o is not a function of x, then R^2 is a constant and the term in the second bracket will represent only a constant phase retardation and can be incorporated into T_0. On the other hand, we also assume that the object wave O is given by Eq. 5.1, but with $O_o = O_o(x)$. This results from the fact that the object is diffuse and the amplitude variation at the hologram plane is just an example of laser speckle. The spatial variation of O_o can be thought of as the self-interference of the light from different parts of the object. Multiplying out the terms in the remaining three brackets and retaining only terms to second order in γ, we obtain

$$T_a(x) = T_o\{1 + i\gamma[|O(x)|^2 + RO^*(x) + R^*O(x)] - \gamma^2 |O(x)|^2$$
$$\times [RO^*(x) + R^*O(x) + |R|^2]$$
$$- \tfrac{1}{2}\gamma^2[|O(x)|^2 + R^2O^*(x)^2 + R^{*2}O^2(x)]\}. \qquad (5.23)$$

The terms in the brackets multiplied by γ are the linear terms that give rise to the desired images. These three terms correspond to, respectively, the flare light around the illuminating beam, the conjugate image, and the primary image. The next set of terms also corresponds to image terms, but since they are multiplied by $\gamma^2O(x)^2$, the flare light that normally only appears around the illuminating beam, now also appears around the desired images. (The noise represented by the third set of terms of Eq. 5.23 does not appear around the images.) This veiling glare is a special characteristic of phase holograms. It is so serious a problem that in spite of the inherently high diffraction efficiencies of phase holograms, the signal-to-noise ratio in the image is often less than that for amplitude holograms.

The problem can be somewhat alleviated by using a high beam balance ratio K defined as the ratio of the average irradiances at the hologram plane of the reference to object beams. However, a high K reduces the diffraction efficiency of the hologram and therefore defeats the whole purpose of producing a phase hologram. For phase holograms recorded on photographic film the situation is even worse because photographic relief images enhance the phase retardation at low spatial frequencies [3], and the spatial variations of the laser speckle $O(x)^2$ are restricted primarily to low spatial frequencies. In the most usual method of making phase holograms, the direct bleach method, this effect is quite pronounced. Phase holograms on thermoplastics, however, probably do not suffer from this form of low-frequency noise because these materials have an inherently low response at low frequencies [2]. However, thermoplastics produce strictly thin holograms and the very high diffraction efficiencies associated with thick phase materials are not available. Dichromated gelatin, on the other hand, is capable of producing holograms of near 100% efficiency, but there is no associated relief image with these materials. At first this may seem advantageous, but it is not. The presence of the relief image can be used to counteract the effects of the index variation at low spatial frequencies, and therefore significantly reduce the flare light. This method of producing phase holograms, called reversal bleach processing, is the only method of producing thick phase holograms with low flare light and high diffraction efficiency. We now describe some of the methods for making phase holograms.

5.3   METHODS FOR PRODUCING PHASE HOLOGRAMS

5.3.1   Direct Bleaching

The simplest method for producing a phase hologram is to bleach out the silver image in a photographic emulsion. This method allows one to use a high-speed panchromatic recording material, and to produce high diffraction efficiency holograms. Unfortunately these holograms also have a large amount of noise associated with them because the direct bleaching process tends to enhance the low spatial frequencies by producing relief images that augment the index images. A typical process is shown in Fig. 5.4. After the film or plate is developed and fixed in the usual manner, it is bleached in some solution that converts the metallic silver into a transparent, insoluble salt having a refractive index significantly higher than that of the gelatin. For example, Kodak bleach bath R-10 converts the silver into silver chloride. (By modifying the B solution it is also possible to produce silver bromide or silver iodide). Many other bleach baths are possible, such as ones containing mercuric chloride $HgCl_2$, cupric bromide $CuBr_2$, ferric chloride $FeCl_3$, ammonium dichromate $(NH_4)_2Cr_2O_7$, or potassium ferricyanide $K_3Fe(CN)_6$.

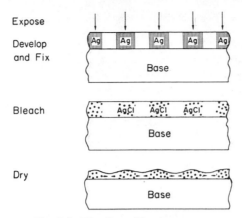

Fig. 5.4 The direct bleach process.

Many workers have described bleach processes using these oxidizing agents. One claimed to be among the best is potassium ferricyanide [9]. Bleaches of this type produce stable, low-noise, high-efficiency holograms. The recommended processing procedure is outlined in Table 5.1.

5.3.2 Dichromated Gelatin [10]

Direct exposure of a gelatin film containing a dichromate, either $(NH_4)_2Cr_2O_7$ or $K_2Cr_2O_7$, will form a phase hologram. Holograms of this type are relatively easy to make and process. The diffraction efficiency is very high (approaching the theoretical 100%) and for the simple two-beam type (i.e., not a diffuse object), they have very low noise. In fact, a properly processed dichromated gelatin hologram is so clear that it looks like a clear glass plate. The disadvantages are that the material is quite insensitive and long exposures are required, and that it is sensitive only to blue light (and uv). A further disadvantage appears when a diffuse object is to be recorded. In this case there is

Table 5.1 Processing Procedure for Direct Bleaching[a]

1. Develop	(Kodak developer D-76)	4 min
2. Wash		30 sec
3. Fix	(Kodak Rapidfixer)	5 min
4. Wash		10 min
5. Bleach	(20 g $K_3Fe(CN)_6$, 10 g NaCl, 1000 cc water)	5 min
6. Wash		10 min
7. Soak	(Formula 30 ethyl alcohol)	3 min
8. Soak	(Ethyl alcohol, 200 proof)	5 min
9. Dry	(Dry N_2 atmosphere)	15 min

[a] All processes at room temperature.

Table 5.2   Processing Procedure for Dichromated Gelatin

1. Soak[a]	$(0.5\% \ (NH_4)_2Cr_2O_7)$	5 min
2. Fix	(Kodak Rapidfixer)	5 min
3. Wash	(Water at $20°C$)	5 min
4. Soak	(Isopropanol)	2 min
5. Dry	(Air)	15 min

[a] The first two steps are to prevent the formation of a milky, cloudy appearance. If this is not a problem these two steps may be eliminated.

no appreciable relief image associated with dichromated gelatin films to compensate for the low spatial frequency noise caused by the self-interference of the object light. As a result the hologram images become noisy because of the veiling glare.

Dichromated gelatin plates are not available commercially and so they have to be prepared by the user. It is not too difficult to prepare them; Lin [10] describes some and gives references to the descriptions of other methods. One particularly simple one consists of simply fixing out the silver halide from a commercial photographic plate. The gelatin film is then washed in water and methanol so that it is free of most chemicals. The sensitization of the film is achieved simply by soaking the gelatin plate in an aqueous solution (5 %) of the dichromate for about 5 min, and then allowing it to dry slowly in the dark. Preparation in this way yields gelatin films that are already hardened. For the other methods of preparation the gelatin film must be hardened before use. This prehardening is the main feature of the method described by Lin; it ensures that no gelatin will be readily dissolved or etched away when the hologram is washed in water. After exposure, the hologram is simply washed in water. The water is absorbed by the gelatin and it expands. The amount of water that is absorbed, and hence the amount of expansion of the gelatin, decreases with increasing exposure. This produces an expansion of the film that corresponds to the image. Upon subsequent rapid dehydration in isopropanol, a hologram is formed both on the surface and in the volume of the film. The exact nature of the chemical processes and the mechanisms of the light diffraction are not well understood; nevertheless, good quality holograms result. The recommended processing procedure is outlined in Table 5.2.

5.3.3   Thermoplastics

The use of a thermoplastic material as a hologram recording material was first discussed by Urbach and Meier [2]. The technique that they proposed and the one that has proved most successful so far consists of a thermoplastic-photoconductor-transparent conductor sandwich that is operated in the

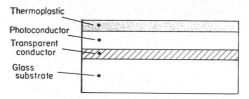

Fig. 5.5   Schematic cross section of a thermoplastic-photoconductor sandwich.

charge-expose-recharge mode. A cross-sectional schematic diagram of the type of sandwich they used is shown in Fig. 5.5. An exposure to an image results in a distribution of static charge on the thermoplastic corresponding to the image. Subsequent heating of the film softens the thermoplastic so that it deforms in accordance with the variation of the electric field corresponding to the image. Because no absorption of light is involved and because the diffraction is caused by the surface relief of the deformed thermoplastic, this system produces strictly thin phase holograms.

Since thin phase holograms can have a maximum diffraction efficiency of 0.33 (see Chapter 7), thermoplastic holograms can have a reasonably large efficiency. This, coupled with the fact that the associated photoconductor is fairly sensitive (100–1000 ergs/cm^2) and has a panchromatic sensitivity, gives us two of the principal advantages of a thermoplastic recording material. The most important advantage, however, is that these materials are reusable. As many as several thousand complete write-read-erase cycles with reasonably good image quality have been reported. A further advantage of the materials stems from the quasiresonant nature of their spatial frequency response [2]. Because of this the DC and low-frequency responses of these materials are quite low, and the low-frequency speckle patterns resulting from the self-interference of light from different parts of a diffuse object are simply not recorded. As a result the flare light usually accompanying phase holograms of diffuse objects is not present.

However, this bandpass or quasiresonant type of spatial frequency response also leads to one of the major disadvantages of this material. The bandwidth is so small (at best around 500 cycles/mm) that it is simply not possible to record a large field of view. This restriction certainly limits the maximum possible information storage capacity. A further disadvantage is that the films are quite difficult to prepare, and to date, at least, preparation must be done by the user since no material of this type is available commercially.

The generally accepted best mode of operation for these materials, according to Urbach and Meier [2], is in the charge-expose-recharge mode, outlined in Fig. 5.6. First the thermoplastic layer is charged by means of a corona discharge to a positive voltage of around 800 V. While in this charged

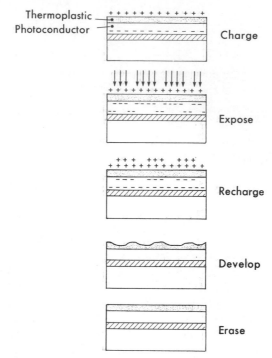

Thermoplastic
Photoconductor

Charge

Expose

Recharge

Develop

Erase

Fig. 5.6   The charge-expose-recharge cycle for thermoplastic phase holograms.

state an exposure causes a transfer of negative charge to the thermoplastic-photoconductor interface, reducing the voltage on the upper surface of the thermoplastic. Next the surface is recharged, returning the surface to a uniform potential. In effect, this increases the electric field existing across the thermoplastic wherever the exposure had caused the photoconductor to discharge. After this step, therefore, there exists a variation of the electric field corresponding to the image. In the final step of the recording process, the thermoplastic layer is heated until it softens. When it is soft, it deforms under the stress of the local electric field, producing, therefore, a thickness variation corresponding to the image. This thickness variation is made permanent by cooling the layer to room temperature.

The important feature of erasure is now accomplished by simply reheating the thermoplastic layer until it softens and holding it there until the soft plastic again flows to a uniform thickness. The material is then ready for a new cycle.

5.3.4 Reversal Bleaching

The reversal bleaching process represents an attempt to overcome one of the principal causes of flare light in the image of extended objects formed with a

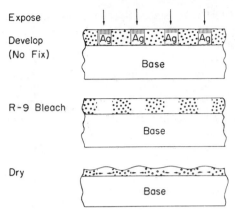

Expose

Develop
(No Fix)

R-9 Bleach

Dry

Fig. 5.7   The reversal bleach process.

phase hologram. As mentioned previously, this cause is the self-interference of the light from different points of the extended object. In most bleaching processes, such as the direct process described above, the surface relief image and variation of refractive index of the bleached emulsion combine to enhance the self-interference pattern at low spatial frequencies and thus to enhance the flare light in the image. With the reverse bleach process, an attempt is made to cause the relief image to cancel the effects of the index variation for low spatial frequencies. This allows simultaneous high diffraction efficiency and signal-to-noise ratio.

To achieve this, the film is first exposed and developed but is not fixed. Thus the silver halide remains in the unexposed regions (see Fig. 5.7). The hologram is then bleached in Kodak bleach bath R-9, which converts the metallic silver to a soluble salt and removes it entirely from the emulsion. Because the remaining silver halide appears reversed from what it is in the direct bleach method, this method is termed the reversal bleach method. The silver halide thus forms a negative image of high index. However, the strong relief at low frequencies occurs where the initial development took place, that is, where maximum tanning action occurred [3]. The relief image thus tends to counteract the phase retardation caused by the high index silver halide. Hence the low-frequency phase information (speckle noise) is effectively canceled out.

The advantages of this system lie in its ability to produce low-noise phase holograms concurrent with a moderately high diffraction efficiency. Since the recording material is photographic film, it is cheap, readily available, very sensitive, and has a panchromatic response. Also, the processing procedure is not too difficult.

One disadvantage of the method is that because of the peaked nature of the relief image response curve [3], there is really only one spatial frequency that is fully suppressed, so that objects of differing angular extents behave

Table 5.3 Kodak Special Developer SD-48

Solution A[a]	
Water	750 ml
Sodium sulfite, desiccated	8 g
Pyrocatechol[b]	40 g
Sodium sulfate, desiccated	100 g
Water to make	1 liter
Solution B	
Water	750 ml
Sodium hydroxide	20 g
Sodium sulfate, desiccated	100 g
Water to make	1 liter

[a] *Caution:* Use rubber gloves or take special precaution to keep hands from coming in contact with the developer.

[b] This is available as Eastman organic chemical P604. It is a hazardous solid material and breathing of dust or vapor can be harmful. Avoid getting pyrocatechol on the skin, in eyes, or on clothing.

differently. The main disadvantage is that the formulation described by Lamberts and Kurtz [8] is really only for one particular emulsion—Kodak Spectroscopic Plate, type 649-F. Because of the different thicknesses of other emulsions, the developer described in Table 5.3 would probably have to be reformulated. In spite of these difficulties, some fairly good results have been achieved with this method. Diffraction efficiencies near 40% with signal-to-noise ratios of nearly 20 have been reported for a small diffuse object. The

Table 5.4 Processing Procedure for Reversal Bleaching

1. Develop[a]	(Kodak Special Developer, SD-48)	5 min
2. Stop	(Kodak Stop Bath SB-1)	15 sec
3. Wash		1 min
4. Bleach	(Kodak Bleach Bath R-9)	3 min
5. Wash		5 min
6. Remove stain	(Kodak Stain Remover S-13, Solution A)	1 min
7. Clear	(Kodak Stain Remover S-13, Solution B)	1 min
8. Wash		5 min
9. Dry[b]		

[a] All processes at 24°C.

[b] For uniform drying, it is suggested that the plate be rinsed in a solution of methanol diluted 1:1 with water, then washed twice in isopropyl alcohol.

Table 5.5   Bleach and Stain Remover Formulas

Kodak Bleach Bath R-9	
Water	1.0 liter
Potassium Dichromate	9.5 g
Sulfuric Acid (concentrated)[a]	12.0 ml
Kodak Stain Remover S-13	
Solution A	
Water	750.0 ml
Sulfuric acid (concentrated)[a]	8.0 ml
Potassium permanganate	2.5 g
Water to make	1.0 liter
Solution B	
Water	750.0 ml
Sodium bisulfate	10.0 g
Water to make	1.0 liter

[a] *Caution:* Always add the sulfuric acid to the solution slowly, stirring constantly, and never the solution to the acid; otherwise the solution may boil and spatter the acid on the hands or face, causing serious burns.

recommended processing procedure is outlined in Table 5.4 and the formulations of the bleach bath and the stain remover are shown in Table 5.5.

5.4   CONCLUSIONS

This concludes our discussion of phase holograms. Some of the materials and processes discussed here are described more fully in Chapter 11 on materials. A much more thorough treatment of diffraction efficiency is given in Chapter 7. There are, of course, many more ways of making phase holograms. There are other systems of thermoplastics and other techniques; there are many other ways of producing phase holograms by bleaching photographic emulsions, by etching processes, by embossing, and so on. We have attempted to discuss here, briefly, only the principal methods by which phase holograms are made today. The future is sure to bring new materials and techniques in the quest for the ideal recording material.

REFERENCES

[1] G. L. Rogers, *Proc. Roy. Soc. Edinburgh*, A63, 193 (1952).
[2] J. C. Urbach and R. W. Meier, *Appl. Opt.*, 5, 666 (1966).

[3] H. M. Smith, *J. Opt. Soc. Am.*, **58**, 533 (1968).

[4] W. T. Cathey, Jr., *J. Opt. Soc. Am.*, **55**, 457 (1965).

[5] A. K. Rigler, *J. Opt. Soc. Am.*, **55**, 1693 (1965).

[6] C. B. Burckhardt, *J. Opt. Soc. Am.*, **57**, 601 (1967).

[7] H. Kogelnik, *Bell Syst. Tech. J.*, **48**, 2909 (1969).

[8] R. L. Lamberts and C. N. Kurtz, *Appl. Opt.*, **10**, 1342 (1971).

[9] C. B. Burckhardt and E. T. Doherty, *Appl. Opt.*, **8**, 2479 (1969).

[10] L. H. Lin, *Appl. Opt.*, **8**, 963 (1969).

6 Color Holography

6.0 INTRODUCTION

Although so far we have discussed only holograms recorded in mono-chromatic light and have not as yet discussed the possibility of superposing more than one hologram in a single recording medium, it should be evident that with only a small extension of the preceding ideas, multicolor wavefront reconstruction should be possible.

The basic idea of color holography is to record three (or more) separate holograms on a single photographic plate (or other suitable recording medium), each with a different color, in such a way that subsequent illumination with a three-color beam yields three separate wavefronts, one in each of the primary colors, representing the portion of the object corresponding to that color.

Multicolor wavefront reconstruction was first proposed by Leith and Upatnieks in one of their original papers [1]. Their technique is the most straightforward of all: the hologram records three incoherently superposed diffraction patterns on a photographic emulsion. Each component hologram is a record of the object as it would appear when illuminated with a single color. If the object and reference beams contain the three primary colors, illumination of the composite hologram with a beam identical to the reference beam yields a reconstructed wavefront that closely approximates the wavefront that would result if the object were illuminated with white light. However, there are several other wavefronts that are produced in the process, and these overlap and generally degrade the color image. These spurious wavefronts, leading to "crosstalk" images, result from light of wavelength λ_1 diffracting from the component hologram recorded with the wavelength λ_2. In this method, there will be six spurious primary images. All of the methods of

91

color holography to be discussed in this chapter represent various schemes to eliminate these crosstalk images. The technique described by Leith and Upatnieks consisted of introducing the reference beam for each primary color at a different angle, thus spatially separating the crosstalk images from the true one.

Later, Mandel [2] noted that the crosstalk images are spatially removed from the desired image by a small amount because of the slightly different angles of diffraction for each color. Thus by viewing the image over a limited field, the reference beams for the three colors need not be introduced at different angles.

Pennington and Lin [3] utilized the thickness of the photographic emulsion to separate the crosstalk images from the desired ones. By using a thick emulsion, the hologram becomes essentially a three-dimensional diffraction grating, diffraction from which is governed by the Bragg relation

$$2d \sin \theta = \lambda, \tag{6.1}$$

where d is the grating spacing, θ the angle of incidence (and diffraction), and λ the wavelength. Figure 4.13 indicates that for small d (large θ_R in Fig. 4.13) and a thick emulsion, only a slight change in λ is required to extinguish the image. Thus for multicolor holograms recorded on a thick emulsion, the cross-talk images are largely suppressed since the illuminating wavelengths do not satisfy Eq. 6.1.

Later, Lin et al. [4] produced two-color holograms of the volume reflection type which yielded two-color images when illuminated with white light. As discussed in Chapter 4, when the object and reference beams enter the recording medium from opposite directions, standing waves are set up that result in a set of reflecting planes lying generally parallel to the emulsion surface. Since these reflecting planes are separated by $\lambda/2$, the hologram acts as a spectral filter, and the crosstalk images will again be suppressed. Because of the filtering action of the hologram, multicolor images may be obtained by illuminating with white light. The hologram will have high reflectance (diffraction) for only those wavelengths used to record the hologram.

Still later, Collier and Pennington introduced two more methods for making multicolor holograms without the requirement of a thick recording medium [5]. The first of these they called spatial multiplexing, in which the hologram is formed in such a way that the various component holograms in each color are not allowed to overlap on the emulsion. In reconstruction, the hologram is illuminated in such a way that each color illuminates only the portion of the hologram corresponding to that color. In this way no crosstalk images are produced.

The second method requires coding of the reference beam so that each component reference wave varies in a unique manner over the hologram plate. In this case the amplitude and phase of the reference wave are made to vary across the hologram plate in a significantly different manner for each of the colors used to form the hologram. To reconstruct, it is necessary to relocate the hologram in exactly the position in which it was recorded, and to illuminate it with a beam identical to the original coded reference beam. In this way the crosstalk images are sufficiently suppressed to render a good three-color image.

These six methods represent the typical schemes for recording multicolor holograms; the list is not meant to be exhaustive. Some of the proposed techniques work independently of the thickness of the recording medium, while others demand that the thickness of the recording medium be large compared to the recorded fringe spacing. The aim behind each technique is to suppress the unwanted crosstalk images.

The requirements for the recording materials for color holography are somewhat different from those for recording monochromatic holograms. First, of course, the recording medium must be sensitive to all of the wavelengths to be used in recording the color hologram. This requirement of panchromatic sensitivity rules out the use of some of the more efficient materials such as dichromated gelatin. Also, it is probably true that some of the most successful color holograms recorded to date have been recorded as volume holograms, and the more complex techniques required to form a color hologram as a plane hologram are not commonly employed. This consideration, for example, rules out the use of any of the strictly thin recording media such as thermoplastics. The best type of recording medium to use for color holography is therefore a thick phase material that has panchromatic sensitivity. Such a material is represented by a thick, bleached photographic emulsion (see Chapter 5). However, most of the color holograms made so far have been made as thick amplitude holograms. The main drawback to this type of hologram is its very low diffraction efficiency. When three-color holograms are recorded as absorption holograms the efficiency drops still farther, as explained in Section 6.4. Holograms that can be viewed in reflection (Section 4.3) can be illuminated with white light during reconstruction, but only a small fraction of this illuminating light is diffracted as the color image, the rest being absorbed (cf. Eq. 4.109). In addition to these problems of low efficiency, there is the problem created by the fact that blue light is much more strongly scattered in a fine-grained recording medium than is red light. This makes it very difficult to record the blue hologram with reasonably high efficiency.

In spite of all of these difficulties, some very successful color holograms have been recorded [4–7]. The techniques used to record these holograms will be described following a brief analysis.

6.1 ANALYSIS

Consider a multicolor wavefront that has been scattered by an object to be given by

$$O(x) = \sum_{i=1}^{n} O_i(x) \tag{6.2}$$

and a multicolor reference wavefront

$$R(x) = \sum_{i=1}^{n} R_i(x). \tag{6.3}$$

Here $R_i(x)$ and $O_i(x)$ represent the complex amplitude at the hologram (x) plane corresponding to the wavelength λ_1.

Since there is no mutual coherence between the waves of different colors, the waves interfere at the hologram plane in such a way as to yield an irradiance

$$|H(x)|^2 = \sum_{i=1}^{n} [|O_i(x)|^2 + |R_i(x)|^2 + O_i^*(x)R_i(x) + O_i(x)R_i^*(x)]. \tag{6.4}$$

We suppose as usual that the resulting amplitude transmittance of the hologram is simply equal to $|H(x)|^2$. Thus when the hologram is illuminated with a multicolor wavefront given by

$$C(x) = \sum_{j=1}^{n} C_j(x) \tag{6.5}$$

the resulting transmitted wave is given by

$$\psi(x) = \sum_{j=1}^{n} C_j(x) \sum_{i=1}^{n} [|O_i|^2 + |R_i|^2 + O_i^*R_i + O_iR_i^*]. \tag{6.6}$$

The first two terms in the brackets on the right represent the n^2 zero orders and n^2 flare terms. The last two terms give rise to n^2 conjugate waves,

$$\sum_{i,j}^{n} C_j O_i^* R_i \tag{6.7}$$

and n^2 primary waves

$$\sum_{i,j}^{n} C_j O_i R_i^*. \tag{6.8}$$

The terms in these sums for which $i \neq j$ yield the unwanted cross-talk images, while the terms for $i = j$ yield the true multicolor images.

To simplify the analysis, let us assume that the object, reference, and illuminating beams originate from point sources. The multicolor object point is located at (x_o, y_o, z_o) and radiates at wavelengths λ_i. The reference point sources are located at $(x_R, y_R, z_R)_i$ and radiate at wavelengths λ_i. The subscript on these position coordinates indicates that each of the n wavelengths of the reference beam may originate from a different point. Similarly, the illuminating wave originates at point sources located at $(x_c, y_c, z_c)_j$ and radiate the wavelengths λ_j.

We may now extend Eq. 9.130 to include multiple wavelengths. By assuming no hologram scaling ($m = 1$), we obtain for the phases of the primary wavefronts (to first order)

$$
\Phi_{P_{ij}}^{(1)} = \frac{2\pi}{\lambda_j} \cdot \frac{1}{2} \left[(x^2 + y^2) \left(\frac{1}{z_{c_j}} + \frac{\mu_{ij}}{z_o} - \frac{\mu_{ij}}{z_{R_j}} \right) \right.
$$
$$
- 2x \left(\frac{x_{c_j}}{z_{c_j}} + \mu_{ij} \frac{x_o}{z_o} - \mu_{ij} \frac{x_{R_j}}{z_{P_j}} \right)
$$
$$
\left. - 2y \left(\frac{y_{c_j}}{z_{c_j}} + \mu_{ij} \frac{y_o}{z_o} - \mu_{ij} \frac{y_{R_j}}{z_{R_j}} \right) \right]. \quad (6.9)
$$

Here $\mu_{ij} = \lambda_j/\lambda_i$ is the ratio of illuminating to reference wavelengths. To simplify the analysis still further, let us omit the y-dependence and just consider the terms in x. Each of the n^2 wavefronts of Eq. 6.9 yield a Gaussian image point located at

$$
Z_{P_{ij}} = \frac{z_{c_j} z_o z_{R_j}}{z_o z_{R_j} + \mu_{ij} z_{c_j} z_{R_j} - \mu_{ij} z_{c_j} z_o} \quad (6.10)
$$

and

$$
a_{P_{ij}} = \frac{x_{c_j} z_o z_{R_j} + \mu_{ij} x_o z_{c_j} z_{R_j} - \mu_{ij} x_R z_{c_j} z_o}{z_o z_{R_j} + \mu_{ij} z_{o_j} z_{R_j} - \mu_{ij} z_{c_j} z_o}. \quad (6.11)
$$

Thus there are in general n^2 Gaussian image points, and the problem is to separate the $n(n - 1)$ unwanted ones from the n desired ones, and moreover, have the n desired points coincident.

To do this, Leith and Upatnieks [1], using red, green, and blue wavelengths, separated the three reference sources so that

$$
(x_R, z_R)_1 \neq (x_R, z_R)_2 \neq (x_R, z_R)_3. \quad (6.12)
$$

Now by illuminating the hologram with an identical wave,

$$
(x_c, z_i)_j = (x_R, z_R)_j, \quad j = 1, 2, 3 \quad (6.13)
$$

they obtained, for example,

$$
Z_{P_{11}} = z_o = Z_{P_{22}} = Z_{P_{33}} \quad (6.14)
$$

and

$$Z_{P_{12}} = \frac{z_{c_2} z_o}{z_o(1 - \mu_{12}) + \mu_{12} z_{c_2}} \tag{6.15}$$

and also

$$a_{P_{11}} = x_o = a_{P_{22}} = a_{P_{33}} \tag{6.16}$$

$$a_{P_{12}} = \frac{x_{c_2} z_o(1 - \mu_{12}) + \mu_{12} x_o z_{c_2}}{z_o(1 - \mu_{12}) + \mu_{12} z_{c_2}}. \tag{6.17}$$

From these examples we see that the mixed terms yield Gaussian image points that are spatially separated from the desired multicolor image point. Obviously it is desirable to have μ_{ij}, $i \neq j$ large so that the cross-talk images are well separated from the true images.

When the thickness of the recording medium is comparable to or greater than the fringe spacing, the amount of diffracted image flux is strongly sensitive to the orientation of the hologram and to the wavelengths of the illuminating beam. With a thick hologram, then, it is much easier to separate the crosstalk images from the true image. Maximum flux is diffracted when the orientation of the three-dimensional fringe system is such as to satisfy Eq. 6.1. The grating spacing d is actually the separation between the planes through the recording medium. These planes act as if they were mirrors to the illuminating light when Eq. 6.1 is satisfied. The production of these planes has been discussed in Chapter 4. If the two interfacing waves contain several spectral components, a set of these reflecting (Bragg planes) surfaces is produced for each component. When this three-dimensional grating is illuminated with a plane wave containing the several spectral components, each component is diffracted according to Eq. 6.1, resulting in a plane wave of the same spectral composition. This is the reconstructed multicolor wavefront. However, there will also be several diffracted waves from, for example, wavelength component λ_1 diffracting from the grating produced by component λ_2. This gives rise to the unwanted crosstalk images. However, for the three-dimensional grating, these waves are much reduced in irradiance, as indicated in Fig. 4.9. Here we see that if the various spectral components are widely separated in wavelength, the recording medium relatively thick, and the grating spacing small, then there will be very little flux diffracted into the unwanted directions. For the case of a more complicated wavefront, the above arguments still apply, since the complicated wavefront can be thought of as a superposition of many plane wavefronts. Thus the three-dimensional hologram is probably the most suitable for multicolor wavefront reconstruction. However, three-color reconstruction is still possible with plane holograms, as described in the following sections.

6.2 PLANE HOLOGRAM TECHNIQUES

6.2.1 Three-Reference-Beam Method

As briefly described in the Introduction, Leith and Upatnieks [1] initially described the simplest and most direct method for producing a three-color image. One simply superposes the three holograms in each of the three primary colors with its reference beam incident on the recording plane at a different angle. During reconstruction, then, assuming that the illumination is by the same three reference beams at the same three angles, the spurious cross-talk images are separated from the desired images because, for example, the red light will diffract from the green light hologram at a different angle than from the red light hologram. Since the desired image results from the red light diffracting from the red hologram and the green light from the green, and so on, the crosstalk images will be spatially separated from the good image because of the different angles of diffraction. This method, however, is not too practical because of the severely restricted field of view that results; therefore, only objects that subtend small angles at the bottom can be successfully recorded using this technique. Another problem with this technique is that three accurately aligned illuminating beams must be used for reconstruction. Since most of the other methods to be described do not have these problems, this method has not gained full acceptance.

6.2.2 Spatial Multiplexing

Another method by which a color hologram can be recorded as a plane hologram has been called nonoverlapping spatial multiplexing [5].

By this method the crosstalk images are completely eliminated by recording the hologram in such a way that the various colored interference patterns are not allowed to overlap on the recording medium. The resulting hologram is thus really a composite of many nonoverlapping component holograms, each recorded in only a single color. To reconstruct the object wavefront, the hologram is illuminated in such a way that each component hologram is illuminated with only that color light used to form it. In this way the crosstalk images are completely suppressed. There will, however, be some loss of resolution with this method since each of the component holograms will be considerably smaller than the total recording, and the total area of, say, the red holograms, will still be less than the total area of the recording.

There are many possible experimental arrangements for producing such a hologram. One possible scheme for recording a three-color wavefront is indicated in Fig. 6.1. The incoming three-color beam is intercepted by a mirror M_1 and a colored, diffusely reflecting object. The reference beam

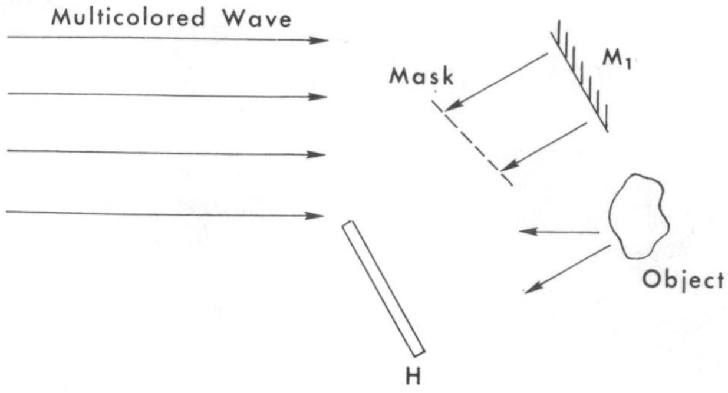

Fig. 6.1 Nonoverlapping spatial multiplexing [5]. M_1 is a mirror and H is the hologram.

coming from the mirror passes through a color filter mask before interfering with the object beam at the hologram plane H. The mask consists of
narrow strips of red, green, and blue filters. Each filter allows only one color
to pass through it, resulting in narrow strips of red, green, and blue light
incident on the hologram plane. The light in each strip then interferes with the
light of the same color coming from the object, yielding a nonoverlapping
multiplicity of single-color holograms. Illumination of the hologram through
this same filter mask reconstructs the object wavefront, with no crosstalk
images. Of course, this scheme results in the production of some noise
because, for example, in a red component hologram there will also be
present green and blue light from the object, which merely decreases the
average transmission of this component hologram. Since the same is true for
each component hologram, the composite hologram will have only a limited
fringe modulation, resulting in a low diffraction efficiency, and hence there
will be a lower signal-to-noise ratio in the image. There are other possible
arrangements that can be used to avoid this drawback. One of these is to place
the color filters directly in contact with the recording material. Each filter thus
intercepts both the object and reference waves and allows only the recording
of a single-color hologram behind it. After processing, of course, and during
reconstruction, the hologram must be placed in exact register with the original
filters. This method offers the possibility of reconstructing with white light, if
the spectral transmission bands of the filters are narrow enough.

6.2.3   Coded Reference Beam

The third method for recording multicolor wavefronts on thin recording
media has been called reference beam coding [5]. With this method, the
amplitude and phase of the reference wave are made to vary across the
hologram plane in a manner that is different for each wavelength. After the
hologram has been exposed and processed, it must be replaced in precisely the

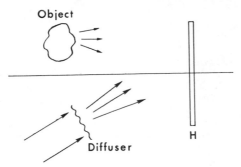

Fig. 6.2   Reference beam coding. The hologram is designated H.

same position it occupied during exposure and illuminated with a beam that is identical in every respect to the original reference beam. If the illuminating beam is not identical to the reference beam, the object wave will not be reconstructed and the viewer will see only a relatively uniform noise distribution.

Figure 6.2 indicates a possible arrangement for recording this type of hologram. The multicolor reference wave is passed through a diffuser, such as a good glass plate. As the wavefront progresses from the coding plate to the hologram plane, each of the color components is dispersed in a different manner. The complicated phase variations imposed on each wave as it passes through the code plate are further scrambled by the dispersion process in a manner that is strongly dependent on the wavelength. Hence each component reference wave is coded in a complicated manner. As an example of the wavelength sensitivity of the process, Collier and Pennington [5] have noted that when only light of wavelength $0.514\,\mu$ was used, an excellent image was obtained. However, by changing the wavelength by only $0.0128\,\mu$ and illuminating the coding plate in exactly the same way, no image at all was obtained.

Successful holograms have been recorded using this method, but there is still a noise problem. The noise results from what would normally produce the crosstalk images, namely, the interaction of wavelengths λ_1 and λ_3 with the hologram formed with wavelength λ_1. The geometry of the code plate strongly affects the degree to which the noise interferes with the image. If the solid angle subtended by the coding plate at the hologram is small, the noise is fairly localized and therefore most annoying. The noise can be distributed more or less uniformly by increasing the solid angle subtended by the code plate.

The resolution obtainable with this method should be quite high, as discussed in Section 9.1. Since the reconstruction process represents an autocorrelation of the reference source distribution, the resolution will be quite high under the same conditions that give the least amount of crosstalk.

6.3   VOLUME HOLOGRAM TECHNIQUES

6.3.1   Transmission Holograms

The first volume color hologram was recorded by Pennington and Lin of the Bell Telephone Laboratories [3]. They reconstructed a two-color wavefront using the arrangement shown in Fig. 6.3. Light of wavelength 0.6328 μ from a He-Ne laser is mixed with blue light wavelength 0.4880 μ from an argon laser at the beamsplitter B. Part of this two-color wavefront proceeds along the upper path and constitutes the reference beam. The remaining part travels the lower path and passes through the object, which in this case was a color transparency. The two beams then interfere and expose the photographic plate P throughout the depth of the emulsion.

Subsequent illumination of the hologram with the blue wavefront alone reconstructed only the wavefronts that the transparency originally transmitted at this wavelength. A similar result was obtained with the red light, indicating that good rejection of the crosstalk images was achieved.

Three-color wavefronts, yielding better color fidelity than the two-color wavefronts, have also been recorded [6]. The basic arrangement used is shown in Fig. 6.4. Two of the spectral components (0.488 μ and 0.514 μ) are derived from an argon laser. These are mixed with a third component at 0.633 μ from a He-Ne laser at the beamsplitter B. The object beam illuminates a diffusely reflecting object, yielding a complex wave of three colors that interfere with the corresponding colors in the reference beam coming from mirror M_2. A

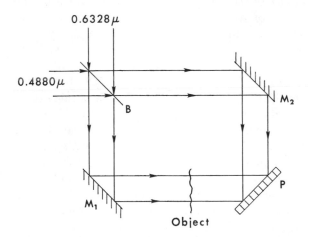

Fig. 6.3   Arrangement used by Pennington and Lin to record a two-color hologram [3]. M_1 and M_2 are mirrors, B a beamsplitter, and P the photographic plate.

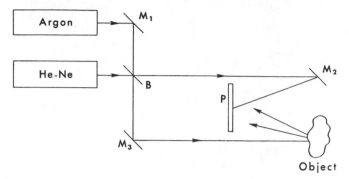

Fig. 6.4   Arrangement for recording a three-color hologram [6].

thick emulsion photographic plate P is exposed to these interference patterns and becomes the hologram. If the angle between the reference and object beams is kept relatively large, the grating spacing is small and crosstalk image rejection good. Better rejection of the crosstalk images can be achieved with a larger separation of the blue and green components.

6.3.2   Reflection Holograms

From Fig. 4.13 we can see that the maximum crosstalk image suppression is achieved when the angle between the multicolor reference and object beams is so large that they enter the recording medium from opposite sides ($\theta_R = 180°$). The Bragg planes then lie approximately parallel to the surface of the recording medium, and the hologram is viewed in reflection. Holograms of this type were discussed in Chapter 4, and the extension of the basic ideas to multicolor wavefront reconstruction is straightforward.

The major feature of this type of hologram is that a multicolor wavefront is reconstructed with white light illumination of the hologram. Since the Bragg planes lie approximately parallel to the emulsion, they are separated by approximately $\lambda/2n$, where n is the index of refraction of the emulsion and λ is the wavelength in air. These planes act as a spectral filter and have a high reflectivity only for those wavelengths that produced them. Hence if the object and reference beams are composed of several wavelength components, there will be several (one for each wavelength) sets of reflecting planes produced. Upon illumination of the resulting hologram with white light, the original spectral components are filtered out. A schematic diagram of the formation and readout of this type of hologram is shown in Fig. 6.5. The multicolored reference beam is incident from one side of the recording medium P, while the object beam enters the other side. Readout is accomplished by placing the eye at E and illuminating the hologram with either white light or a beam identical to the reference beam. A full-color image of the

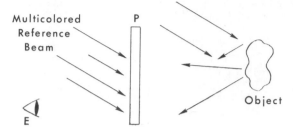

Fig. 6.5   Recording a multicolor hologram that can be viewed in reflection by illumination with white light.

object can be observed. Illumination of the hologram with a beam is problematical if the hologram has been recorded on a photographic emulsion. Because of emulsion shrinkage, the Bragg planes pull together so that the final spacing is no longer $\lambda/2n$. Instead the spacing is $\lambda'/2n$, with $\lambda' < \lambda$, so that little or no flux is reflected at the original wavelengths. Because of this, white light illumination is often more suitable, unless steps have been taken to prevent emulsion shrinkage or the hologram has been recorded on some other recording medium. In fact, for normal processing of normal photographic emulsions, the shrinkage is great enough ($\sim$15–20%) that the blue component will be shifted entirely out of the visible spectrum. Obviously, no color fidelity at all will be achieved unless steps are taken to reduce the shrinkage and the hologram is illuminated with white light. White light illumination does, however, result in a loss of image resolution, since the hologram acts as a spectral filter with a bandwidth of typically 100 Å or so, which is too large for high-resolution imagery.

6.4   DIFFRACTION EFFICIENCY

When recording a color hologram, the overall diffraction efficiency drops drastically compared to a single-color hologram because each of the color components is recorded independently of the others, that is, incoherent superposition is used. If there are N such components to be recorded (usually three for color holograms), then the overall efficiency decreases by a factor of $1/N^2$ compared to a single-color recording. To see how this comes about, consider N object points to be recorded. (In the case of a color hologram these can be considered to be each of a different color so that they are mutually incoherent.) Suppose first of all that all the object points are illuminated with a single color so that they radiate coherently. If the amplitude at the hologram caused by each point (assumed equal) is $O_n e^{i\varphi_n}$ and that due to the reference beam is R_o, then the beam ratio K is

$$K = \frac{R_o^2}{NO_n^2} \tag{6.18}$$

and the total exposure is

$$E_{tot} = \left| \sum_{n=1}^{N} O_n e^{i\varphi_n} + R_o \right|^2$$

$$= NO_n^2 + R_o^2 + 2O_n R_o \sum_{n=1}^{N} \cos \varphi_n \tag{6.19}$$

since the φ_n are considered to be all equal. Each of the cosine terms leads to the image of the corresponding object point. The modulation of one of these terms is

$$M_c = \frac{2O_n R_o}{NO_n^2 + R_o^2} \tag{6.20}$$

where the subscript c refers to coherent superposition. The efficiency with which one of these terms (one of the object waves) will be diffracted is proportional to M_c^2.

Next, suppose that the light from each of the object points is recorded incoherently, that is, either sequentially or in different colors. To maintain the same exposure and beam ratio as in the coherent case, the reference and object beam irradiances at the hologram must be reduced by the factor N. Hence the exposure for each of the object points is

$$E_n = \left| O_n e^{i\varphi_n} + \frac{R_o}{N^{1/2}} \right|^2$$

$$= O_n^2 + \frac{R_o^2}{N} + \frac{2O_n R_o}{N^{1/2}} \cos \varphi_n \tag{6.21}$$

and the total exposure is

$$E_{tot} = \sum_{n=1}^{N} E_n = N \left(O_n^2 + \frac{R_o^2}{N} \right) + \frac{2O_n R_o}{N^{1/2}} \sum_{n=1}^{N} \cos \varphi_n. \tag{6.22}$$

The modulation of one of the image components in this case is

$$M_{inc} = \frac{2O_n R_o / N^{1/2}}{NO_n^2 + R_o^2} = \frac{M_c}{N^{1/2}}. \tag{6.23}$$

The amount of light diffracted into one of the images in the single-color (coherent) case is simply proportional to the reconstructing light (R_o^2) times the square of the modulation,

$$\text{Diffracted light} \propto R_o^2 M_c^2. \tag{6.24}$$

For the case of a color hologram, only a fraction R_o^2/N of the reconstructing light will diffract from the hologram component with modulation M_{inc} so that in this case

$$\text{Diffracted light} \propto \frac{R_o^2}{N} M_{inc}^2 = \frac{R_o^2 M_c^2}{N^2}, \tag{6.25}$$

which is seen to be only $1/N^2$ as much light per image point as in the coherent single-color case. Since the diffraction efficiency of volume holograms (see Section 7.2) is also proportional to the square of the modulation, this reduction in efficiency obtains no matter what type of color hologram is recorded, thick or thin. There is a slight difference, however, for the case of a spatially multiplexed color hologram. This case does not really represent incoherent superposition since a given area of the hologram records in only a single color. In this case there is only the reduction in the reconstructing light to consider. For N colors (assuming equal areas), only $1/N$ of the total beam will be used for reconstruction, thus allowing only $1/N$ of the possible diffracted light. This reduction in diffraction efficiency for color holograms seems to be an unfortunate necessity. Attempts to improve the situation with bleaching techniques have so far not been very successful.

6.5   COLORIMETRY

The colors that can be reproduced holographically depend upon the choice of the three primary colors. These in turn depend upon the choice of lasers to be used. Not all lasers are suitable for color holography because of considerations of power at the desired wavelength and coherence length. For the sake of convenience, a power of 5–10 mW should be available at the desired wavelengths. This keeps the exposure times down to the order of a few minutes with the commonly used Kodak Spectroscopic Plate, Type 649F, even for relatively large objects.

The required coherence length is determined by the size of the object and the size of the hologram to be recorded. The most common choice of lasers for color holography today is the He-Ne laser with emission at 0.633 μ and the argon laser with lines at 0.488 μ and 0.514 μ. All of these lines can be produced with ample power, but the argon laser has a relatively short ($\sim$8 cm) coherence length. Various ways have been proposed for increasing the coherence length [4] and single-mode oscillation at 0.514 μ has been attained [8].

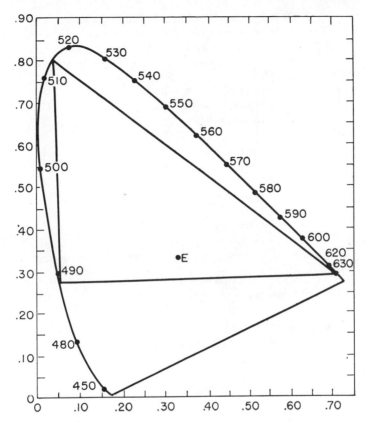

Fig. 6.6 The standard CIE chromaticity diagram. The point E represents the chromaticity of a source radiating equal flux per unit bandwidth throughout the visible spectrum. The colors within the triangle can be reproduced with mixtures of the three wavelengths 0.488, 0.514, and 0.633 μ.

The colors which can be reproduced with these three wavelengths can be determined by plotting them on the standard C.I.E. chromaticity diagram, shown in Fig. 6.6. This diagram gives the relative proportion x of one primary as the abscissa and that of another y as the ordinate. All visible colors are contained within the boundaries defined as the spectrum locus, consisting of the monochromatic colors from 0.700 μ down to 0.400 μ, and the line of purples, extending from 0.400 μ to 0.700 μ.

When any two colors are mixed, the new color lies somewhere along the line joining the two mixed colors in the diagram. Since the shape of the diagram is essentially triangular, one can see that three suitably chosen colors can be mixed in various proportions so as to yield a large range of colors, but not all. For the wavelengths 0.488, 0.514, and 0.633 μ used for most color work, any

Table 6.1 Laser Wavelengths for Holography

Wavelength (Å)	Laser	Power (mW)	Coherence Length (cm)
4416	He-Cd	50	4
4579	Argon	50	8
4762	Krypton	25	12
4765	Argon	250	8
4880	Argon	1000	8
4965	Argon	150	8
5017	Argon	150	8
5145	Argon	1000	8
5208	Krypton	50	12
5682	Krypton	50	12
6328	He-He	25	20
6471	Krypton	150	12

color within the triangle indicated in Fig. 6.6 can be reproduced. It is seen that the deep blues and saturated (near the spectrum locus) yellows and greens will not be reproduced.

There are, however, several other laser lines that could prove to be more suitable for color holography. These are listed in Table 6.1.

REFERENCES

[1] E. N. Leith and J. Upatnieks, *J. Opt. Soc. Am.*, **54**, 1295 (1964).

[2] L. Mandel, *J. Opt. Soc. Am.*, **55**, 1697 (1965).

[3] K. S. Pennington and L. S. Lin, *Appl. Phys. Letters*, **7**, 56 (1965).

[4] L. S. Lin, K. S. Pennington, G. W. Stroke, and A. E. Labeyrie, *Bell Syst. Tech. J.*, **45**, 659 (1966).

[5] R. J. Collier and K. S. Pennington, *Appl. Opt.*, **6**, 1091 (1967).

[6] A. A. Friesem and R. J. Fedorowicz, *Appl. Opt.*, **6**, 529 (1967).

[7] L. H. Lin and C. V. LoBianco, *Appl. Opt.*, **6**, 1255 (1967).

[8] J. M. Forsyth, *Appl. Phys. Letters*, **11**, 391 (1967).

7 Diffraction Efficiency

7.0 INTRODUCTION

Diffraction efficiency is defined as the ratio of the total, useful, image forming light flux diffracted by the hologram to the total light flux used to illuminate the hologram. If some of the illuminating light is diffracted into an unwanted conjugate (or primary) image, it is still only the light in the primary (or conjugate) image that is pertinent to the stated diffraction efficiency. The importance of producing a hologram with a reasonable diffraction efficiency is obvious, whether the application is simply the viewing of a three-dimensional scene, holographic interferometry, or data storage: efficient use of the illuminating light allows for smaller and more economical systems. Being able to use a smaller and cheaper laser for illumination or readout, for example, could be the deciding factor in determining the feasibility of a holographic data storage system.

However, one must approach the question of diffraction efficiency with some careful thought. In many cases it is the output signal-to-noise ratio (s/n) that is the more important consideration. For example, a great many papers have been published describing various means for achieving high diffraction efficiency, mostly various photographic schemes, *without any consideration being given to the buildup of noise in the image*. All of this work represents many thousands of hours of research effort for the development of systems that will never be used because of intolerable noise levels, in spite of very high efficiency.

The noise in a holographic image arises principally from three factors: intrinsic nonlinearity, such as in phase holograms, that causes unwanted light to be diffracted into the image region; the nonlinearity of the amplitude transmittance-exposure characteristic of the recording medium; and the noise caused by the buildup of the granular structure of the recording material.

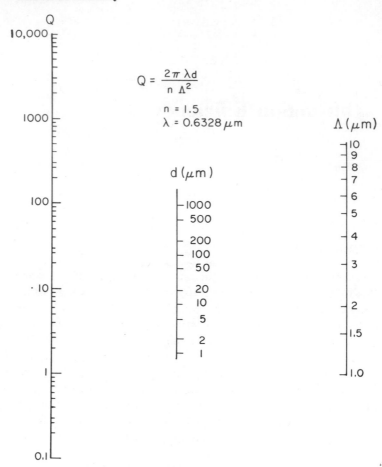

Fig. 7.1 Nomograph relating Q, hologram thickness d, and fringe spacing Λ for holograms recorded with a wavelength 0.6328 μ and having an index of 1.5.

A complete discussion of the sources of noise in holographic images follows in Chapter 8. With this warning in mind, we note that it is nevertheless possible to increase the diffraction efficiency by various techniques without seriously impairing the s/n characteristics of the recording.

The equations describing the diffraction efficiency of the various hologram types are derived in the following sections. The sections are divided into two groups, plane holograms and volume holograms. This is because the description of the diffraction process for the two cases is fundamentally different, even though it will be shown (Section 7.3) that in the common case of amplitude holograms, for practical recording situations, the resulting diffraction efficiencies are equivalent. This is definitely not the case for phase holograms, however. The distinction between thick and thin holograms (volume or plane) is usually made with the aid of the Q-parameter defined as

$$Q \equiv \frac{2\pi\lambda d}{n\Lambda^2}. \tag{7.1}$$

The hologram (or grating) is considered to be thick when $Q \geq 10$ and thin otherwise. A nomograph relating Q to the illuminating wavelength λ, the hologram thickness d, and the fringe spacing Λ, and an index of refraction n of 1.5, is shown in Fig. 7.1.

7.1 PLANE HOLOGRAMS

7.1.1 Amplitude Holograms

We begin by assuming an irradiance distribution at the hologram (for a diffuse object) given by

$$\begin{aligned} H(x) &= |O_o(x)e^{i\varphi_o(x)} + R_o e^{i\omega x}|^2 \\ &= O_o^2(x) + R_o^2 + 2O_o(x)R_o \cos(\omega x - \varphi_o), \end{aligned} \tag{7.2}$$

where $O_o(x)$ and R_o are the (real) amplitudes and $\varphi_o(x)$ and ωx are the phases of the object and reference waves, respectively. The object is assumed to be diffuse so that both the phase and amplitude of the object wave are functions of the space coordinate x. The average spatial frequency of this distribution is given by

$$\bar{\nu} = \frac{1}{2\pi}\left\langle \frac{d}{dx}[\omega x - \varphi_o(x)] \right\rangle, \tag{7.3}$$

where the angle brackets indicate a spatial average. If we take into account the modulation transfer function (MTF) of the emulsion, the effective exposure is

$$E = tH = t(\langle O_o^2(x)\rangle + R_o^2)\{1 + mM(\bar{\nu})\cos(\omega x - \varphi_o)\}, \tag{7.4}$$

where t is the exposure time and m is the exposure modulation given by

$$m = \frac{2(\langle O_o^2(x)\rangle R_o^2)^{1/2}}{\langle O_o^2(x)\rangle + R_o^2}. \tag{7.5}$$

In Eq. 7.4 we have used $\langle O_o^2(x)\rangle$ rather than the more rigorous $O_o^2(x)$. Hence we are neglecting the self-interference or speckle caused by the diffuse object. However, the effects of the speckle are not excessive as long as the modulation m is not too large. The modulation of the fringes represented by the cosine term of Eq. 7.4 is reduced by the factor $M(\bar{\nu})$, the MTF of the emulsion at the average frequency $\bar{\nu}$. This is a good approximation as long as the object size is such that it subtends a very small angle at the hologram. If this is the case, $\varphi_o(x)$ is nearly a constant, or a constant plus a term proportional to x, and the bandwidth of the recorded signal is small and centered around $\omega/2\pi$. If we assume the amplitude transmittance-log exposure curve (T_a-log E) to be essentially linear with slope α in the vicinity of the average exposure $E_o = t(\langle O_o^2(x) \rangle + R_o^2)$, we can write for the amplitude transmittance

$$T_a(E) = T_a(E_o) + \alpha \log \frac{E}{E_o}, \tag{7.6}$$

so that

$$T_a(x) = T_a(E_o) + \alpha \log \{1 + mM(\bar{\nu}) \cos (\omega x - \varphi_o)\}. \tag{7.7}$$

Expanding the logarithmic function we have

$$T_a(x) = T_a(E_o) + 0.434\alpha \left[mM(\bar{\nu}) \cos (\omega x - \varphi_o) - \frac{m^2 M^2(\bar{\nu})}{2} \cos^2 (\omega x - \varphi_o) \right.$$
$$\left. + \frac{m^3 M^3(\bar{\nu})}{3} \cos^3 (\omega x - \varphi_o) - \cdots \right]. \tag{7.8}$$

The T_a-log E curve is linear only in the vicinity of E_o, which implies a moderately small exposure modulation m. Therefore, we must assume that m is small enough that we can neglect all except the first-order term in Eq. 7.8 and arrive at

$$T_a(x) = T_a(E_o) + 0.434\alpha m M(\bar{\nu}) \cos (\omega x - \varphi_o). \tag{7.9}$$

The object wave is reconstructed by illuminating the hologram with the reference wave $R_o e^{i\omega x}$, yielding a conjugate image term

$$\psi_c(x) = \frac{0.434}{2} R_0 \alpha m M(\bar{\nu}) e^{i\varphi_o}. \tag{7.10}$$

The irradiance in the reconstructed field is

$$H_c = R_o^2 \left[\frac{0.434}{2} \alpha m M(\bar{\nu}) \right]^2 \tag{7.11}$$

and the diffraction efficiency is

$$\eta = \frac{H_c}{R_o^2} = \left[\frac{0.434}{2} \alpha m M(\bar{\nu}) \right]^2. \tag{7.12}$$

Now

$$m = \frac{2(\langle O_o^2(x) \rangle R_o^2)^{1/2}}{(\langle O_o^2(x) \rangle + R_o^2)} = \frac{2\sqrt{K}}{1 + K}, \tag{7.13}$$

where K is the beam ratio $R_o^2/\langle O_o^2(x)\rangle$. Since we are already restricted to moderately small values of m, K is somewhat greater than unity, and so

$$m \approx \frac{2}{\sqrt{K}} \qquad (7.14)$$

giving for the diffraction efficiency

$$\eta = 0.188\, \frac{\alpha^2}{K}\, M^2(\bar{\nu}). \qquad (7.15)$$

This equation shows that the diffraction efficiency will be a maximum at an average exposure for which the T_a-log E curve has maximum slope. The slope of this curve is related to the slope γ of the conventional D-log E curve through the equation

$$\alpha(E) = -1.15\, T_a(E)\gamma(E), \qquad (7.16)$$

where $\gamma(E) = dD/d\log E$.

Because we have assumed that the exposure modulation m is not too large, Eq. 7.15 does not indicate the maximum diffraction efficiency obtainable with thin amplitude holograms. To get an idea of this upper limit, we write for the amplitude transmittance of the exposed and processed hologram

$$T_a(x) = T_a(E_o) + T_1 \cos(\omega x - \varphi_o) \qquad (7.17)$$

where $T_a(E_o)$ is the average transmittance due to the average exposure

$$E_o = t(\langle O_o^2(x)\rangle + R_o^2)$$

and T_1 is the amplitude of the spatially varying portion of the exposure,

$$t(\langle O_n^2(x)\rangle + R_n^2)mM(\bar{\nu}) \cos(\omega x - \varphi_o).$$

To achieve the maximum possible diffraction efficiency, $T_a(x)$ should vary between 0 and 1. Thus

$$
\begin{aligned}
T_a(x) &= \tfrac{1}{2} + \tfrac{1}{2}\cos(\omega x - \varphi_o) \\
&= \tfrac{1}{2} + \tfrac{1}{4}e^{i(\omega x - \varphi_o)} + \tfrac{1}{4}e^{-i(\omega x - \varphi_o)}.
\end{aligned} \qquad (7.18)
$$

From this equation it is evident that the maximum diffracted amplitude in either the primary or conjugate image is only one-fourth the amplitude of the illuminating wave. The useful light flux in either image is therefore only one-sixteenth of the illuminating flux, yielding a maximum diffraction efficiency of 0.0625. This is somewhat artificial, however, since there are really no practical materials that are linear over an exposure range so large that $T_a(x)$ will vary from 0 to 1. Some nonlinearity or clipping will always occur when the exposure modulation gets large. Therefore, the maximum efficiency of 0.0625 cannot be achieved when it is necessary that the reconstructed wave amplitude be proportional to the object wave amplitude.

On the other hand, if some nonlinearity can be tolerated, somewhat higher diffraction efficiencies can be obtained. For example, if $T_a(x)$ varies as a square wave, such as a Ronchi grating, the transmittance is given by

$$T_a(x) = \frac{1}{2} + \frac{2}{\pi} \cos{(\omega x - \varphi_o)} - \cdots$$

$$= \frac{1}{2} + \frac{1}{\pi} [e^{i(\omega x - \varphi_o)} + e^{-i(\omega x - \varphi_o)}] - \cdots. \qquad (7.19)$$

In this case a fraction $(1/\pi)^2$ of the illuminating flux is diffracted into either of the two first-order images, giving a diffraction efficiency of 0.101. There will now, of course, be some light diffracted into the higher orders also, and some nonlinearity noise will be present. However, this type of hologram could result from a computer-generated hologram, for example, wherein this analysis could apply.

7.1.2   Phase Holograms

Referring back to Eq. 5.13, we see that the transmission term leading to the primary image in a phase hologram is given by

$$T_p(x) = T_o \exp{i(\gamma O_o^2 + \gamma R_o^2 + \pi/2)} J_1(2\gamma O_o R_o) \exp{i(\varphi_o - \beta x)} \qquad (7.20)$$

where T_o is the average transmittance, γ is the proportionality constant between exposure and phase modulation (see Eq. 5.3), O_o and R_o are the real amplitudes of the object and reference waves, respectively, and $J_1(x)$ is the Bessel function of the first kind and first order. When the hologram is illuminated with a wave of unit amplitude, the first-order diffracted wave has an amplitude T_p given by Eq. 7.19. Since we are assuming for simplicity that we have a pure phase hologram that is, no absorption, $T_0 = 1.0$ and the diffracted wave amplitude is simple $J_1(2\gamma O_o R_o)$. The diffraction efficiency is equal to the square of this quantity. The maximum value possible is 0.339, which is considerably higher than the maximum efficiency for amplitude holograms. It will be seen from the following sections that the efficiency of thick phase grating is even higher yet, which explains the very great interest in producing low-noise phase holograms.

7.2   VOLUME HOLOGRAMS

7.2.1   Coupled-Wave Theory

The preceding theories for thin holograms cannot apply when the diffraction efficiency becomes high, because for high diffraction efficiencies the illuminating wave will be strongly depleted as it passes through the grating. In some way one must take into account the fact that at some point within the grating there will be two mutually coherent waves of comparable magnitude traveling together. Such an account is the basis of the coupled-wave theory, so aptly applied to the problem of diffraction from thick holograms by Kogelnik [1]. This elegant analysis gives closed-form results for the angular and wavelength sensitivities for all of the possible hologram types—transmission and reflection, amplitude or phase, with and without loss, and with slanted or unslanted fringe planes. The equations also give, of course, the maximum diffraction efficiency achievable when the grating is illuminated at the Bragg angle and the fringe planes are unslanted. The equations are the result of a theory that assumes that the gratings are relatively thick so that there are only two waves in the medium to be considered, that is, that the Bragg effects are rather strong. Nevertheless, the equations are surprisingly accurate over a very large range of Q-values, including values considerably less than 10 [2].

For a really complete treatment of the theory one should refer to [1]. Here we will only outline briefly the underlying ideas of the theory and give the principal results.

The coupled-wave theory assumes that there are only two waves present in the grating: the illuminating wave R and the diffracted signal wave S. It is assumed that the Bragg condition is approximately satisfied by these two waves and that all other orders strongly violate the Bragg condition and hence are not present. The equations that are derived express the coherent interaction between the waves R and S.

Figure 7.2 defines the grating assumed by Kogelnik for his analysis. The z-axis is perpendicular to the surfaces of the medium, the x-axis is in the plane of incidence and parallel to the medium boundaries, and the y-axis is perpendicular to the page. The fringe planes are oriented perpendicularly to

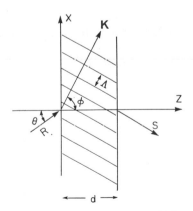

Fig. 7.2   Notation used to define a thick grating.

the plane of incidence and slanted with respect to the medium boundaries at an angle φ. The grating vector $\mathbf{K}$ is oriented perpendicularly to the fringe planes and is of length $|\mathbf{K}| = 2\pi/\Lambda$, where Λ is the period of the grating. The angle of incidence for the illuminating wave R in the medium is θ. The Bragg angle is θ_o. It is assumed that the fringes are sinusoidal variations of the index of refraction or of the absorption constant, or both for the case of mixed gratings. The assumed equations are therefore

$$n_o = n + n_1 \cos (\mathbf{K} \cdot \mathbf{X}) \tag{7.21}$$

and

$$a_o = a + a_1 \cos (\mathbf{K} \cdot \mathbf{X}) \tag{7.22}$$

where $\mathbf{K} \cdot \mathbf{X} = \omega x$ and ω is 2π divided by the fringe spacing in the x-direction. The quantities n and a are the average values of the index and absorption constant, respectively. The slant of the fringe planes is described by the constants c_R and c_s, which are given by

$$c_R = \cos \theta \tag{7.23}$$

$$c_s = \cos \theta - \frac{K}{\beta} \cos \theta \tag{7.24}$$

where $\beta = 2\pi n/\lambda$ and λ is the free space wavelength.

When the Bragg condition is satisfied we have

$$\cos (\varphi - \theta_o) = \frac{K}{2\beta} \tag{7.25}$$

and

$$c_R = \cos \theta_o \tag{7.26}$$

$$c_s = -\cos (2\varphi - \theta_o). \tag{7.27}$$

In the case where the Bragg condition is not satisfied, either by a deviation $\Delta\theta$ from the Bragg angle θ_o or by a deviation $\Delta\lambda$ from the correct wavelength λ_o, where both $\Delta\theta$ and $\Delta\lambda$ are assumed to be small, Kogelnik introduces a dephasing measure ϑ. This parameter is a measure of the rate at which the illuminating wave and the diffracted wave get out of phase, resulting in a destructive interaction between them. The dephasing measure is given by

$$\vartheta = \Delta\theta \cdot K \sin (\varphi - \theta_o) - \frac{\Delta\lambda \cdot K^2}{4\pi n} \tag{7.28}$$

where

$$\Delta\theta = \theta - \theta_o \tag{7.29}$$

and

$$\Delta\lambda = \lambda - \lambda_o. \tag{7.30}$$

The coupled-wave equations that result from this theory are

$$c_R R' + \alpha R = -i\kappa S \qquad (7.31)$$

$$c_s S' + (a + i\vartheta)S = -i\kappa R \qquad (7.32)$$

where κ is the constant defined by

$$\kappa = \frac{\pi n_1}{\lambda} - \frac{i a_1}{2}, \qquad (7.33)$$

and the primes denote differentiation with respect to z, and $i = \sqrt{-1}$. The physical interpretation of these equations as described by Kogelnik is as follows:

As the illuminating and diffracted waves travel through the grating in the z-direction their amplitudes are changing. These changes are caused by the absorption of the medium as described by the terms a_R and a_S, or by the coupling of the waves through the terms κS and κR. When the Bragg condition is not satisfied, the dephasing measure ϑ is nonzero and the two waves become out of phase and interact destructively.

The solutions to these equations take on various forms depending on the type of hologram (grating) considered. When transmission holograms are considered, we say that the fringe planes are unslanted when they are perpendicular to the surface. This condition is described by $c_R = c_s = \cos\theta$ since $\varphi = \pi/2$ in this case. If the Bragg condition is also satisfied, then of course $\theta = \theta_o$. For reflection holograms, on the other hand, no slant means that the fringe planes are parallel to the surface, so $\varphi = 0$. If the Bragg condition also holds, then $c_R = -c_s \cos\theta_o$.

7.2.2 Transmission Holograms

7.2.2.1 Pure Phase Holograms

For a pure phase, transmission hologram the absorption constant $a_o = 0$ and the solution of the coupled-wave equations leads to a diffraction efficiency, for the general case of slanted fringes and Bragg condition not satisfied, given by

$$\eta = \frac{\sin^2 (v^2 + \zeta^2)^{1/2}}{(1 + \eta^2/v^2)}$$

$$v = \frac{\pi n_1 d}{\lambda (c_R c_s)^{1/2}} \qquad (7.34)$$

$$\zeta = \frac{\vartheta d}{2c_s} = \frac{\Delta\theta K d \sin(\varphi - \theta_o)}{2c_s} = -\frac{\Delta\lambda K^2 d}{8\pi n c_s}.$$

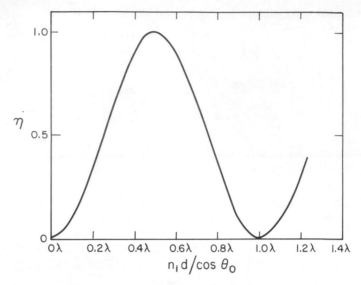

Fig. 7.3   Diffraction efficiency η as a function of the optical path variation $n_1 d/\cos \theta_o$ in units of the wavelength for a thick phase hologram viewed in transmission.

In the case for which there is no slant and the Bragg condition is satisfied, the formula reduces to the well-known equation

$$\eta = \sin^2 \left\{ \frac{\pi n_1 d}{\lambda \cos \theta_o} \right\}. \tag{7.35}$$

A graph of this equation as a function of the optical path $n_1 d/\cos \theta_o$ in units of the free-space wavelengths λ is shown in Fig. 7.3. It is seen that diffraction efficiencies of 1.0 are possible with this type of hologram.

7.2.2.2   Pure Phase Holograms with Loss

For this case we consider the effect of some loss on a phase hologram. The loss takes the form of some residual absorption, such as might be caused by a chemical stain that has not been washed out or perhaps some intrinsic absorption at the readout wavelength. For this sort of grating, a_1 is zero but a is not. The equation for the diffraction efficiency takes the form

$$\eta = \frac{e^{-2ad/\cos \theta} \sin^2 (\nu^2 + \zeta^2)^{1/2}}{(1 + \zeta^2/\nu^2)^{1/2}}$$

$$\nu = \frac{\pi n_1 d}{\lambda \cos \theta} \tag{7.36}$$

$$\zeta = \frac{\vartheta d}{2 \cos \theta} = \Delta \theta \beta d \sin \theta_o$$

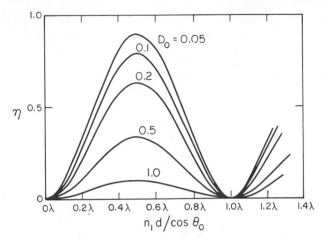

Fig. 7.4 Diffraction efficiency η as a function of the optical path variation $n_1d/\cos\theta_o$ in units of the wavelength for a thick phase hologram with loss, for several values of optical density D_o.

where we assume that the grating is unslanted and that the Bragg condition is not fulfilled. The absorption is evidenced by the exponential term in front and leads to an overall decrease in the efficiency. The angular sensitivity, that is, the sensitivity of the efficiency to Bragg mismatch, is not much different from that of the lossless hologram. In the case where the Bragg condition is fulfilled, the equation simplifies to

$$\eta = e^{-2ad/\cos\theta_o}\sin^2\frac{\pi n_1 d}{\lambda\cos\theta_o}. \qquad (7.37)$$

The effect of the absorption is simply to decrease the overall efficiency by the exponential factor in front. A curve of this function is shown in Fig. 7.4 for several values of the residual density $D_o = 0.868ad/\cos\theta_o$, which is just the optical density measured in the direction of the illuminating wave.

7.2.2.3 *Amplitude Holograms*

For a pure amplitude hologram there is no modulation of the refractive index, so $n_1 = 0$. The solution to the coupled-wave equations in this case is

$$\eta = \frac{c_R}{c_s}\exp\left[-ad\left(\frac{1}{c_R}+\frac{1}{c_s}\right)\right]\sinh^2(\nu^2+\zeta^2)^{1/2}\bigg/\left(1+\frac{\zeta^2}{\nu^2}\right)$$

$$\nu = \frac{a_1 d}{2(c_R c_s)^{1/2}} \qquad (7.38)$$

$$\zeta = \tfrac{1}{2}ad\left(\frac{1}{c_R}-\frac{1}{c_s}\right)$$

for the case where the Bragg condition is fulfilled but the fringe planes are slanted. For the case where the fringes are unslanted but the Bragg condition is not satisfied, the equation for the diffraction becomes

$$\eta = \frac{e^{-2ad/c_R} \sinh^2 (\nu^2 - \zeta^2)}{(1 - \zeta^2/\nu^2)}$$

$$\nu = \frac{a_1 d}{2 \cos \theta}$$

$$\zeta = \frac{\vartheta d}{2 \cos \theta} \simeq \Delta\theta \beta d \sin \theta_o \tag{7.39}$$

$$= -\frac{1}{2} \frac{\Delta\lambda}{\lambda} Kd \tan \theta_o$$

and in case the Bragg condition is satisfied this simplifies to

$$\eta = e^{-2ad/\cos \theta_o} \sinh^2 [a_1 d/2 \cos \theta_o] \tag{7.40}$$

which is plotted in Fig. 7.5. The maximum diffraction efficiency is achieved when $a_1 = a$ and $ad/\cos \theta_o = \ln 3 = 1.1$, and is 0.037. This is considerably lower than the theoretical maximum for a thin amplitude hologram (0.0625). The maximum is reached when $ad/\cos \theta_o = 1.1$, which corresponds to an average density of 0.955 (measured in the direction θ_o) or an amplitude transmittance of 0.34. More will be said about the relationships between thick and thin holograms in Section 7.3.

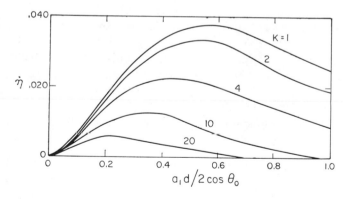

Fig. 7.5   Diffraction efficiency η as a function of the parameter $a_1 d/2 \cos \theta_o$ for thick amplitude holograms viewed in transmission. The curves are plotted for several values of the reference-to-object beam ratio K, which is related to the absorption constants a and θ through $a/a_1 = 2\sqrt{K}/(1 + K)$. The average specular transmittance of the hologram is $e^{-2ad/\cos \theta_o}$.

7.2.2.4 Mixed Holograms

A mixed hologram is one for which there is a phase modulation in addition to the amplitude modulation. In this case both a_1 and n_1 are nonzero. This situation is of some practical importance because any real hologram will be, more or less, a mixed hologram. For example, for a grating produced on a photographic emulsion there is quite likely to be an exposure-dependent variation in the refractive index or a surface relief image in addition to the transmittance variation. This then would result in a mixed hologram. The relative contribution of the phase and amplitude portions can be predicted from coupled-wave theory. Interestingly enough, it turns out that the contributions of the phase and amplitude parts are simply additive, at least for the case of an unslanted grating and Bragg incidence. The equation for the diffraction efficiency is

$$\eta = \left[\sin^2 \frac{\pi n_1 d}{\lambda \cos \theta_o} + \sinh^2 \frac{a_1 d}{2 \cos \theta_o} \right] e^{-2ad/\cos \theta_o} \tag{7.41}$$

where all of the symbols have been previously defined.

7.2.3 Reflection Holograms

7.2.3.1 Pure Phase Holograms

The general equation for the diffraction efficiency of a pure phase reflection hologram with slanted fringe planes and non-Bragg illumination is

$$\eta = \left\{ 1 + \frac{1 + \zeta^2/\nu^2}{\sinh^2 (\nu^2 - \zeta^2)^{1/2}} \right\}^{-1}$$

$$\nu = \frac{i \pi n_1 d}{\lambda (c_R c_s)^{1/2}}$$

$$\zeta = -\frac{\vartheta d}{2 c_s} \tag{7.42}$$

$$= \frac{\Delta \theta K d \sin (\theta_o - \varphi)}{2 c_s}$$

$$= \frac{\Delta \lambda K^2 d}{8 \pi n c_s} .$$

Note that ν is real since c_s is negative. For an unslanted grating and Bragg incidence Eq. 7.41 becomes

$$\eta = \tanh^2 \frac{\pi n_1 d}{\lambda \cos \theta_o} . \tag{7.43}$$

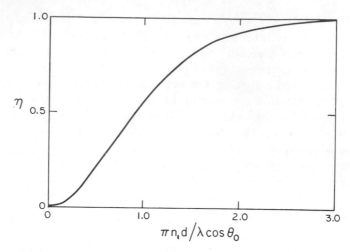

Fig. 7.6 Diffraction efficiency η as a function of the parameter $\pi n_1 d / \lambda \cos \theta_o$ for a pure phase reflection hologram.

Figure 7.6 shows a plot of this equation. It is seen that it is possible to have a diffraction efficiency of 1.0.

7.2.3.2 *Amplitude Holograms*

For a reflection hologram of the amplitude type, where $n_1 = 0$ and a_1 is non-zero, the diffraction efficiency for unslanted fringes ($\varphi = 0$) and Bragg incidence is given by

$$\eta = \frac{c_R}{c_s} \left\{ \frac{\zeta}{\nu} + \left(\frac{\zeta^2}{\nu^2} - 1 \right)^{1/2} \coth (\zeta - \nu^2)^{1/2} \right\}^{-2}$$

$$\nu = \frac{i a_1 d}{2(c_R c_s)^{1/2}}$$

$$\zeta = D_o - i \zeta_o \tag{7.44}$$

$$D_o = \frac{ad}{\cos \theta_o}$$

$$\zeta_o = \Delta \theta \beta d \sin \theta_o = \frac{1}{2} \frac{\Delta \lambda}{\lambda} K d.$$

If the Bragg condition is satisfied, then this equation simplifies to

$$\eta = \frac{D_1^2}{4[D_o^2 + (D_o^2 - D_1^2/4)^{1/2} \coth (D_o^2 - D_1^2/4)^{1/2}]^2} \tag{7.45}$$

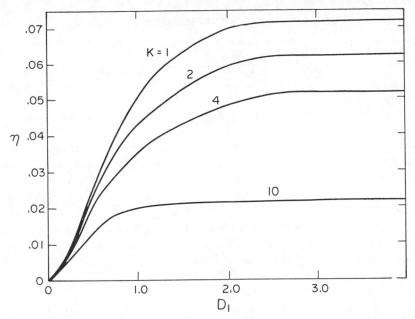

Fig. 7.7 Diffraction efficiency η as a function of $D_1 = a_1 d/\cos\theta_o$ (the specular density) for several values of the beam ratio K, which is related to the densities D_0 and D_1 through $D_1/D_0 = 2\sqrt{K}/(1 + K)$.

where $D_1 = a_1 d/\cos\theta_o$. This equation is plotted in Fig. 7.7. The largest allowable modulation is for $D_1 = D_o$ or equivalently, $a_1 = a$. In this case the maximum diffraction efficiency is 0.072, which is achieved in the limiting case $D_1 = D_o = \infty$.

7.2.4 Summary

Figure 7.8 summarizes the principal results of this section. The numbers shown are the maximum theoretical diffraction efficiency obtainable for each type of hologram. With the exception of the thick phase, reflection hologram, each of these theoretical maxima have been very nearly achieved experimentally. For a reason that is as yet unexplained, no holographic material and process has yet come close to the theoretical 1.0 diffraction efficiency for this type of hologram. Clearly the highest efficiencies are achieved for phase-type holograms, and these can be formed quite readily on many types of holographic recording materials. Unfortunately, as described in Chapter 5, this type of hologram is also inherently noisy, but there are techniques for reducing this noise. Also, for each of the hologram types considered, it was assumed that some sort of linear relationship existed between input and output. For the

THEORETICAL MAXIMUM DIFFRACTION EFFICIENCY

Hologram Type:	Thin Transmission		Thick Transmission		Thick Reflection	
Modulation :	Amplitude	Phase	Amplitude	Phase	Amplitude	Phase
Efficiency :	0.0625	0.339	0.037	1.000	0.072	1.000

Fig. 7.8 Table of maximum diffraction efficiency for the various hologram types.

case of the thick holograms it was between the amplitude transmittance and the exposure. Unfortunately, it is impossible in any real system to maintain this linearity over the exposure range required to produce maximum efficiency. Thus in most real situations one has to balance an increase in nonlinearity noise against an increase in diffraction efficiency. According to Kogelnik [3], the efficiencies that are possible while maintaining the assumed linear response are smaller than the corresponding maximum efficiencies by about 50%, depending on how strictly linearity is defined.

7.3 EQUIVALENCE OF THICK AND THIN HOLOGRAMS

There has been a great deal of experimental work demonstrating the accuracy of Eq. 7.15 for the diffraction efficiency of thin holograms. The same is true of Eq. 7.40 for thick holograms. However, in the case of the experimental verifications of Eq. 7.15, the holograms were recorded on photographic emulsions at relatively high spatial frequencies, so that in a strict sense they were not really thin holograms. It turns out that the reason that both of these equations give accurate results for both thick and somewhat less thick holograms is that they are essentially equivalent, except for cases of high modulation recordings [4]. The essential equivalence of thick and thin holograms is true only for amplitude holograms and not for phase holograms. For the case of phase holograms the thickness has a very significant effect and the diffraction efficiencies are substantially different.

To show the equivalence for the case of amplitude holograms, we begin with Eq. 7.40, for illumination at the Bragg angle θ_o:

$$\eta = e^{-2a_1 d/\cos\theta_o}\sinh^2\frac{a_1 d}{2\cos\theta_o}. \qquad (7.46)$$

As before, a and a_1 are the average and amplitude of the assumed sinusoidal modulation of the absorption coefficient:

$$a_o = a + a_1 \cos\omega x. \qquad (7.47)$$

The density, measured in the direction θ_o, is related to the absorption coefficient by

$$D_o = \frac{0.868 a_o d}{\cos \theta_o} \tag{7.48}$$

so that in terms of density we have

$$\eta = e^{-2.3D} \sinh^2 (0.575 D_1) \tag{7.49}$$

where D and D_1 are the average and amplitude of the sinusoidally varying density. They are obtained by combining Eqs. 7.47 and 7.48 with the result

$$D = \frac{0.868 a d}{\cos \theta_o} \tag{7.50}$$

and

$$D_1 = \frac{0.868 a_1 d}{\cos \theta_o} . \tag{7.51}$$

Expanding the sinh² function we have

$$\eta = e^{-2.3D}(0.575 D_1 + 0.032 D_1^2 + \cdots)^2$$
$$= e^{-2.3D}(0.33 D_1^2 + 0.037 D_1^4 + \cdots). \tag{7.52}$$

The ratio of the second to the first term of this expansion is

$$\frac{0.037 D_1^4}{0.33 D_1^2} = 0.11 D_1^2. \tag{7.53}$$

Now the maximum diffraction efficiency for amplitude holograms occurs for $D_o \approx 0.8$, and D_1 will therefore be smaller. Hence if we neglect all but the first term in the expansion (7.52) the error in the calculated diffraction efficiency is less than 10%. Doing this, we obtain

$$\eta = 0.33 e^{-2.3D} D_1^2 \tag{7.54}$$

We note that the assumption of a sinusoidal absorption coefficient (Eq. 7.47) implies that

$$D_o = D + D_1 \cos \omega x \tag{7.55}$$

because D_o is linearly related to a_o according to Eq. 7.48. Since this density variation arises from an exposure variation in the form

$$E = E_o(1 + mM \cos \omega x) \tag{7.56}$$

we must have

$$D_o = gE + C, \tag{7.57}$$

where C is a constant and g is the slope of the D-E curve. Hence by making use of Eqs. 7.16 and 7.57 we can write for the slope α of the T_a-log E curve

$$\alpha(E_o) \equiv \frac{dT_a}{d \log E}\bigg|_{E_o} = -1.15 e^{-1.15 D_o} \frac{dD_o}{dE} \frac{dE}{d \log E}\bigg|_{E_o}$$

$$= -2.66 e^{-1.15 D} E_o \tag{7.58}$$

where we have also used the definitions

$$T_a = e^{-1.15 D_o} \tag{7.59}$$

and

$$D_o(E_o) = gE_o + C = D. \tag{7.60}$$

It follows from Eqs. 7.56 and 7.57 that $D_1 = gE_o mM$, so by using Eq. 7.58 we have

$$D_1^2 = (gE_o mM)^2 = 0.14 \alpha^2 m^2 M^2 e^{2.3 D}. \tag{7.61}$$

Substituting Eq. 7.61 into Eq. 7.54 and using Eq. 7.14, we have

$$\eta = 0.188 \frac{\alpha^2}{K} M^2(\bar{\nu}) \tag{7.62}$$

which is identical to Eq. 7.15 derived for the thin hologram case. The conclusion to be drawn here is that when the hologram is illuminated at the Bragg angle the thickness dependence drops out, unless, of course, the modulation m is exceptionally large. The hologram, even though it may be thick, behaves as if it were thin.

REFERENCES

[1] H. Kogelnik, *Bell Syst. Tech. J.*, **48**, 2909 (1969).

[2] F. G. Kaspar, *J. Opt. Soc. Am.*, **63**, 37 (1973).

[3] H. Kogelnik, *Proceedings of the Symposium on Modern Optics*, Polytechnic Institute of Brooklyn, 1967, p. 605.

[4] H. M. Smith, *J. Opt. Soc. Am.*, **62**, 802 (1972).

8 Noise

8.0 INTRODUCTION

The noise associated with a holographic image is any unwanted light that is scattered or diffracted into the direction of the desired image during the reconstruction process. This noise light can take the form of a veiling glare that reduces the contrast of the image, or it can be in the form of an unwanted, spurious image. All of the forms of noise present in holography are signal dependent in that they depend either on the amount or form of the object light; because of the coherent nature of the reconstruction process, the noise light adds coherently to the signal or image light.

The principal sources of noise light are as follows:

1. Granularity. When a photographic emulsion is used as the recording material, the information recorded on the hologram is built up from millions of tiny clumps of silver called grains. These photographic grains appear microscopically as fine black specks on a clear background. When light is transmitted through this pattern it is scattered over a wide range of directions, and this is the light that forms a veiling glare around the desired holographic image.

There are some recording materials that are either grainless or essentially grainless, and hence do not scatter light. A thermoplastic material, for example, utilizes the surface relief to diffract the light and the material itself is grainless, and so one would expect no noise of this type. The photochromic materials have grains that are molecular in size. These are so small that the material is essentially grainless and scatters no light.

2. Scattering from the support and/or binder material. Scattering can arise even in the grainless materials because of the optical imperfections of the host material. The photochromics are crystalline materials that are subject to imperfections that scatter light. In the case of the thermoplastics it would be the plastic itself plus the photoconductor layer plus the support material that would scatter the light. In the case of the photographic types of materials one often has to consider the scattering from the support material and from the gelatin binder itself.

3. *Phase noise.*   Phase noise is intrinsic to phase holograms, but occurs only in the case of diffuse objects. The self-interference of the object light is recorded as a low-frequency pattern on the hologram. This low-frequency phase pattern diffracts unwanted light into and around the holographic image.

4. *Nonlinearity noise.*   Nonlinearity noise arises in cases where the final amplitude transmittance of the hologram is not strictly proportional to the recording exposure. The presence of recorded information on the hologram that is proportional to the second and higher orders of the exposure gives rise to unwanted light diffracted in the image direction. This type of noise occurs mainly when the exposure modulation is high so that the range of exposures exceeds the dynamic range of the recording material.

5. *Speckle noise.*   There is another type of noise associated with holographic images, but it does not strictly fit the definition given in the first paragraph above. This is speckle noise. In a strict sense it is not really noise because it arises precisely because of an accurate reconstruction of the object wavefront. However, it is noise in the sense that the image is definitely degraded by its presence. It arises whenever a diffuse wavefront of coherent light is limited in extent by the aperture of the system. When this happens the phase undulations of the diffuse wavefront become amplitude variations, which have a granular or salt-and-pepper appearance. There are many ways to reduce or control this noise and they will be discussed below.

8.1 GRANULARITY

Scattering of light from silver grains is a source of noise that can become important in some situations, such as when one is attempting to store a great deal of information on a single hologram so that the light diffracted into a single image point is very weak. In this case the signal light and the scattered noise light can become comparable in magnitude. When this happens the presence or absence of a particular data point becomes uncertain and errors can result. Severe granularity can also be troublesome when one is simply making pictorial or display holograms, by causing a background or veiling glare around the image that is not pleasing to view.

Because of the nature of holographic imaging, the most convenient characterization of granularity of a holographic emulsion is the noise-power or

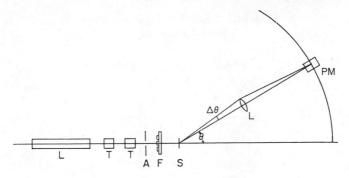

Fig. 8.1 Schematic diagram of the arrangement for measuring scattered light. The laser is denoted by L, beam expanding telescopes by T, aperture A, filter F, sample S, lens L, and photomultiplier PM.

Wiener spectrum. The Wiener spectrum is defined as the modulus squared of the Fourier transform of the transmittance of emulsion. In this case it is the amplitude transmittance that is referred to, and not the flux transmittance. For this reason the noise spectrum of a granular material can be measured simply with a system such as shown in Fig. 8.1. The sample is uniformly exposed and processed to some average amplitude transmittance $\bar{T}_a$. When the sample is illuminated with a monochromatic plane wave, light is scattered over a wide range of angles. If a detector with a small aperture is placed a long way from the illuminated sample, that is, in the far field, the recorded signal is proportional to the modulus squared of the transmittance fluctuations of the sample, which is simply the Wiener spectrum, or noise-power spectrum, of the sample. The angle θ that the detector makes with the axis of the system determines the spatial frequency component of the transmittance variations that is being measured. In the horizontal plane the relationship is

$$\nu = \frac{\sin \theta}{\lambda} \tag{8.1}$$

where ν is the spatial frequency on the sample and λ is the illuminating wavelength. A similar relationship holds in the vertical plane, although because of the isotropic nature of the scattered radiation, measurements are usually made only in the horizontal plane. As the detector scans through a range of angles θ, then, a corresponding range of spatial frequencies is being scanned. Table 8.1 lists the angles and corresponding spatial frequencies for a wavelength of 632.8 nm. This is the wavelength at which most of the measurements of this sort have been made.

Because of the nonzero size of the collecting aperture on the detector, the measurement is actually being made of a range of spatial frequencies, a two-dimensional spatial frequency bandwidth centered on the frequency given by

Table 8.1   Scattering Angle for Given Spatial Frequency

Cycles/mm	Degrees
0	0
100	3°38′
200	7°16′
300	10°57′
400	14°40′
500	18°27′
600	22°19′
700	26°18′
800	30°25′
900	34°43′
1000	39°15′
1100	40°7′
1200	49°24′
1300	55°21′
1400	62°22′
1500	71°40′

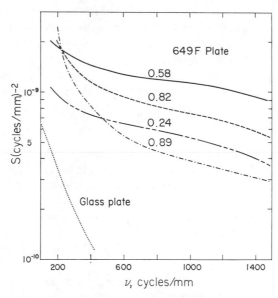

Fig. 8.2 Scattering ratio S as a function of spatial frequency for Kodak Spectroscopic plate, Type 649-F, at several values of amplitude transmittance.

Eq. 8.1. Thus the data are usually presented as a curve of scattered flux per unit two-dimensional bandwidth, per unit incident flux, as a function of spatial frequency on the sample, or equivalently, as a function of angle.

Several measurements of this type have been made [1–7]. A typical result is shown in Fig. 8.2 for Kodak Spectroscopic Plate, type 649F. It is seen from these measurements that at any fixed spatial frequency (angle) the scattered flux is highest at an intermediate level of amplitude transmittance and low at both high and low values of the transmittance. A curve of scattered flux as a function of amplitude transmittance is shown in Fig. 8.3, and indicates a definite peak at $T_a = 0.6$. This is just the position of the peak that is predicted theoretically from the random dot model of the photographic emulsion [7].

Figure 8.4 shows how the noise spectrum increases with increasing development time. This type of data is important because it is well known that the diffraction efficiency of most photographic recording materials increases with increasing development time. However, as is demonstrated by these curves, the signal-to-noise ratio in the image may not improve with increasing development time.

Figure 8.5 shows how the amount of scattered light varies with the grain size of the emulsion. As would be expected, the scattering increases as the

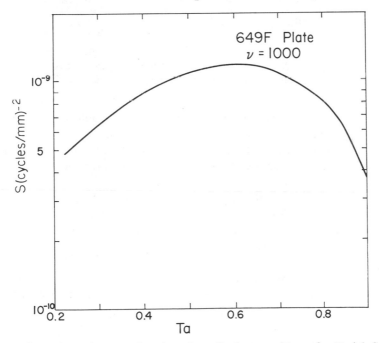

Fig. 8.3 Scattering ratio S as a function of amplitude transmittance for Kodak Spectroscopic plate, Type 649-F at a spatial frequency of 1000 cycles/mm.

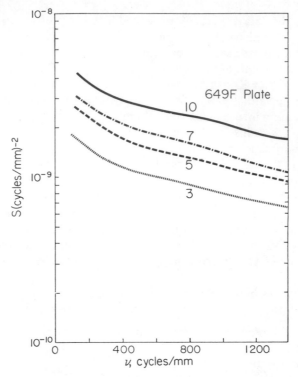

Fig. 8.4 Scattering ratio S as a function of spatial frequency for Kodak Spectroscopic plate, Type 649-F, developed in Kodak HRP developer diluted 1:4 for the indicated times at 21°C.

emulsion grain size increases. This is the reason that even if the requirements of a certain holographic system are such that a relatively coarse-grained, high-speed photographic emulsion could be used, it may not be desirable to do so because the extra scattered light would be objectionable.

This brings up an important point: Just how does one measure how deleterious the scattered light is to an image? Goodman [8] has calculated the signal-to-noise ratio (SNR) for holographic imaging based on a random dot model for the photographic emulsion. Because the scattered flux in a coherent optical system is coherent with the signal flux, the two fields will interfere and a careful definition of SNR must be made. Goodman showed that in the presence of an average scattered irradiance $\bar{I}_N$ from a multitude of randomly phased amplitudes, the SNR is given by the ratio of the deterministic image irradiance I_i to the standard deviation σ ($\sigma^2 \equiv [\overline{(I - \bar{I})^2}]$) of the total irradiance,

$$\frac{I_i}{\sigma} = (I_i/\bar{I}_N)[1 + (2I_i/\bar{I}_N)]^{-1/2} \tag{8.2}$$

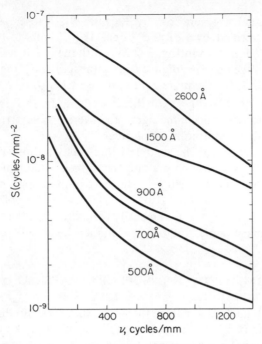

Fig. 8.5 Scattering ratio S as a function of spatial frequency for emulsions of various grain size. The amplitude transmittance of the samples was about 0.6.

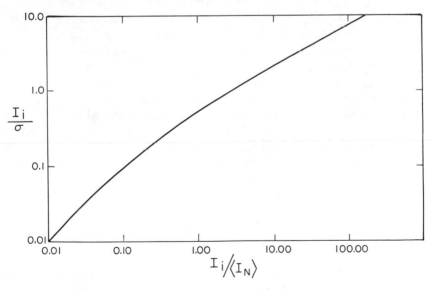

Fig. 8.6 Signal-to-noise ratio I_i/σ as a function of $I_i/\bar{I}_N$.

This result was later generalized by Kozma [9], whose results have been experimentally verified by Lee and Greer [10]. Signal-to-noise ratio as a function of $I_i/\bar{I}_N$ is shown in Fig. 8.6. When the image irradiance I_i is much less than the average noise irradiance $\bar{I}_N$ (an unusual case), then SNR $\approx I_i/\bar{I}_N$. On the other hand, for the more common case when I_i is much greater than $\bar{I}_N$, then SNR $= (1/\sqrt{2})[I_i/\bar{I}_N]^{1/2}$. In this case, then, the SNR is proportional to the ratio of the square roots of the signal and average noise irradiances, that is, to the ratio of signal and noise amplitudes.

By using the curves for the scattered flux such as those in Figs. 8.2–8.5, which give the flux scattered into a unit two-dimensional bandwidth, one can compute, knowing the geometry of the particular holographic system, the average noise irradiance $\bar{I}_N$. Also, with a knowledge of the diffraction efficiency of the hologram, one can easily compute what the signal irradiance will be at the image plane, and hence compute the SNR.

8.2 SCATTERING FROM THE SUPPORT MATERIAL

Aside from the light scattered from the grains that make up the image, there is also the problem of light being scattered from the support material itself. In the case of the very fine-grained emulsions, this is the predominant source of scattered light. The SNR in the image is calculated as described in the preceding section. Curves for the scattered flux spectrum are shown in Fig. 8.7 for three common support materials. Comparison with Fig. 8.5 shows that for grain sizes below about 900 Å, the scattering from the support dominates the grain scattering (except for glass). For the larger grain sizes the scattering from the grains predominates.

These curves indicate only a few of the support materials that can contribute to the scattering. Gelatin itself scatters a small amount of light, and so this will be a consideration for materials such as dichromated gelatin. For the photoconductor-thermoplastic type of materials, the scattering from the photoconductor and from the thermoplastic must be considered. For the photochromic materials it is the scattering from the host crystal that is important. Each system has its own unique scattering properties to be considered.

8.3 PHASE NOISE

There are two distinct kinds of noise that can be classified as phase noise. Both occur only in phase holograms and are a result of the nature of the diffraction of light from a phase object. The first of these noise types arises

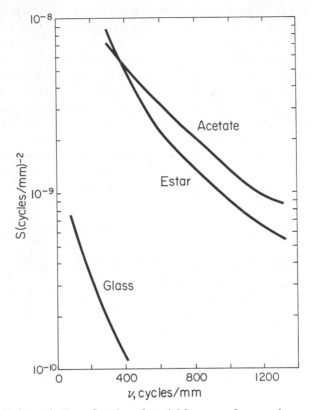

Fig. 8.7 Scattering ratio S as a function of spatial frequency for several support materials.

only for the case in which the object is diffuse; its nature and source have already been treated in Section 5.2. Because a diffuse object scatters light in all directions, light from different parts of the object will mutually interfere at the hologram, resulting in the familiar speckle pattern. In the case of a phase hologram, this pattern will be recorded as a phase image whose appearance has been aptly described as "showerglass." Because of the nature of phase holograms (see Section 5.2), one must view the desired first-order image through this showerglass, which tends to scatter unwanted light into the image and to distort the image.

The other type of phase noise is a result of the intrinsic nonlinearity of phase holograms. As an example of this, consider the simple case in which the final phase modulation to be imposed on the illuminating wave is simply proportional to the exposure $E(x)$, say $\varphi(x) = \gamma E(x)$ as in Eq. 5.3. Then the amplitude transmittance is given by Eq. 5.5 as

$$T_a(x) = T_o e^{i\gamma E(x)}. \tag{8.3}$$

Most of the effect of this nonlinear relation between amplitude transmittance and exposure is to produce diffracted orders higher than the first. These orders are diffracted at angles that are two or more times the angle of the desired first-order image, and thus are usually well separated from it. The existence of higher-order images is generally not important as long as they do not overlap the first-order images. The criterion for prevention of overlap is that the spatial bandwidth of the holographic signal modulating the carrier should be less than one-third the carrier frequency.

What we are concerned with here as noise are the additional disturbance terms in the total reconstructed wavefront which propogate along the first-order image-forming bundle and can thus degrade the image quality. To demonstrate this effect we expand Eq. 8.3 as follows:

$$T_a(x) = T_o e^{i\gamma E(x)} = T_o \left[1 + i\gamma E + \frac{1}{2}(i\gamma E)^2 + \frac{1}{3!}(i\gamma E)^3 + \cdots \right]. \quad (8.4)$$

Substituting $|O + R|^2$ (Eq. 5.20) for $E(x)$ and grouping terms according to diffracted orders gives the transmittance as

$$
\begin{aligned}
\frac{T_a(x)}{T_o} &= 1 + i\gamma[|O|^2 + |R|^2] \\
&\quad - \tfrac{1}{2}\gamma^2[(|O|^2 + |R|^2)^2 + 2|O|^2|R|^2] \\
&\quad - \tfrac{1}{6}\gamma^3[(|O|^2 + |R|^2)^3 + 6|O|^4|R|^2 + 6|O|^2|R|^4 + \cdots]
\end{aligned}
\right\} \quad \text{0 order}
$$

$$
+ OR^*\left[i\gamma - \gamma^2|O|^2 - \gamma^2|R|^2 - \frac{i}{2}\gamma^3|O|^2|R|^2 + \cdots \right] \quad +1 \text{ order}
$$

$$
+ O^*R\left[i\gamma - \gamma^2|O|^2 - \gamma^2|R|^2 - \frac{i}{2}\gamma^3|O|^2|R|^2 + \cdots \right] \quad -1 \text{ order}
$$

$$
+ O^2R^{*2}\left[-\frac{\gamma^2}{2} - \frac{i}{2}\gamma^3(|O|^2 + |R|^2) + \cdots \right] \quad +2 \text{ order}
$$

$$
+ O^{*2}R^2\left[-\frac{\gamma^2}{2} - \frac{i}{2}\gamma^3(|O|^2 + |R|^2) + \cdots \right] \quad -2 \text{ order}
$$

$$
- \frac{i}{6}\gamma^3 O^3 R^{*3} + \cdots \quad +3 \text{ order}
$$

$$
- \frac{i}{6}\gamma^3 O^{*3}R^3 + \cdots. \quad -3 \text{ order}
$$

$$(8.5)$$

Since the reconstructed wave is obtained by multiplication of the above amplitude transmittance by the illuminating wave, the reconstructed wave has the corresponding terms. All terms containing only absolute squares of O and R constitute the zero-order wave. The terms containing OR^* and O^*R constitute the two first-order images. It can be seen that the first-order images contain disturbance terms proportional to $|O|^2$, which affect image quality in a manner that depends on the structure of the object, and on the proportionality constant γ.

Hence we see that the intrinsic nonlinearity of phase holograms (Eq. 8.3) results in degradation of the holographic image. This degradation depends not only on the specific form of the nonlinearity but also on the object itself. While the image of a single point object will not be disturbed by nonlinear effects, which merely result in creation of higher order images, other types of objects generally will be adversely affected.

8.4 NONLINEARITY NOISE

The same sort of nonlinear noise that we have just been discussing can also be present in amplitude holograms if the relationship between exposure and amplitude transmittance is not linear. In this case it is also possible to get some spurious light diffracted into the desired image.

To see how this occurs, we first assume that the nonlinear relationship between amplitude transmittance and exposure can be expressed as a power series of the form

$$T_a = a_0 + a_1 E + a_2 E^2 + a_3 E^3 + \cdots. \tag{8.6}$$

In almost all real cases the strongest nonlinear term is the quadratic term with coefficient a_2, so we will concern ourselves with this term only. In any case this treatment will serve to demonstrate the origin of this sort of noise.

Considering only the term quadratic in E, then, we have

$$E = |O|^2 + |R|^2 + OR^* + O^*R \tag{8.7}$$

so that

$$
\begin{aligned}
a_2 E^2 = a_2 (&|O|^4 + |R|^4 + 4\,|O|^2\,|R|^2 + 2O\,|O|^2\,R^* \\
&+ 2O^*\,|O|^2\,R + 2R^*\,|R|^2\,O + 2R\,|R|^2\,O^* \\
&+ O^2 R^{*2} + O^{*2} R^2) \\
\equiv\ &T_2, \tag{8.8}
\end{aligned}
$$

where this equation serves to define T_2, the component of the transmittance resulting from the quadratic nonlinearity. Assume that R is a plane wave making an angle θ with the hologram so that

$$R = R_o e^{ikx \sin \theta}$$
$$R^* = R_o e^{-ikx \sin \theta} \tag{8.9}$$
$$RR^* = R_o^2 = \text{const.}$$

The reconstructed wave has a nonlinear component RT_2 if the illuminating wave is exactly the same as the reference wave. The only terms that we will consider are those that contribute light in the direction of the image. These terms will not have either R or R^* as a factor. We see that the terms $2O\,|O|^2\,R^*$ and $2R^*\,|R|^2\,O$ will, when multiplied by R, not have R or R^* as a factor. These terms give rise to light diffracted in the direction of the image. However, the latter term is simply proportional to the object wave and contributes linearly to the image. This term boosts the diffraction efficiency but does not lead to any distortion of the image. This leaves only the term $2O\,|O|^2\,R^*$ as providing noise in an amplitude hologram (remembering, of course, that we are here dealing only with the quadratic nonlinearity). The other terms give rise to ghost images and scattered light but generally do not overlap the image.

The term $2O\,|O|^2\,R^*$ can be particularly troublesome when the object is a set of discrete points. In this case false images can be produced by this term, and these can give misleading results in the case of binary coding [11].

8.5 SPECKLE NOISE

Speckle noise is different from the above sources because it does not arise from any non-ideality of the recording material itself. Rather it arises because the holographic imaging method creates such a precise duplicate of the object wave. Since the holographic images of diffuse objects are so faithful, they suffer from the same defect as any image of a diffuse object formed with coherent light, namely, that the observer sees a granular pattern in the image, giving the diffuse surface a scintillating, salt-and-pepper appearance. This whole general problem is known as *laser speckle*. A typical example of laser speckle is shown in Fig. 8.8.

The size of the laser speckles depends on the resolving power of the imaging system in the case where the diffuse surface is being imaged, and on the size of the illuminated spot and the distance to the viewing plane in the case where no imaging system is involved. When an imaging system is involved, the size of the speckle is equal to the size of the least resolved spot on the diffuse surface. Hence as the aperture of the imaging system is stopped down more and more, the resolution becomes less and less and the speckle size becomes greater.

Fig. 8.8 Examples of laser speckle in a holographic image. The photographs are taken at various f/numbers to produce the different amounts of speckle.

Speckle arises because of the scattering of coherent light from a diffuse surface. Each scattering center of the diffuse surface acts like a point source of monochromatic light that is mutually coherent with all of the other point sources on the diffuse surface. Each, however, differs in phase from all of the others in a random fashion. The light from all of these points within one barely resolvable spot of the diffuse surface is brought to a focus within an area the size of the spread function of the imaging system. Depending on how the phases add up within one of these elementary areas, it can range from completely dark to a bright spot.

In the case where there is no imaging system involved, speckle is still formed by the same process, except that in this case the aperture is simply the area of the diffuse surface that is illuminated. The size of the speckles depends on the ratio of the diameter of this area to the distance to the plane on which the speckles are observed.

The RMS fluctuations of irradiance in a speckle pattern can be calculated as follows. Assume that each scattering center produces an amplitude a_k in the image plane, and each has a phase φ_k. The phases are assumed to be random, with any value between 0 and 2π equally likely. Within a just resolvable area in the image plane the irradiance is

$$H = \left| \sum_{k=1}^{N} a_k e^{i\varphi_k} \right|^2 = \left| a_T e^{i\varphi_T} \right|^2 = a_T^2 \tag{8.10}$$

where N is the number of scattering centers within a just resolvable area. The average irradiance is

$$\bar{H} = \overline{\left| \sum_{k=1}^{N} a_k e^{i\varphi_k} \right|^2} = \overline{a_T^{\,2}}. \tag{8.11}$$

We want to calculate the RMS fluctuations in this quantity, namely

$$\sigma = [\overline{(H - \bar{H})^2}]^{1/2} = [\overline{H^2} - \bar{H}^2]^{1/2}. \tag{8.12}$$

The problem of finding the sum

$$\sum_{k=1}^{N} a_k e^{i\varphi_k} = a_T e^{i\varphi_T}$$

is analogous to the famous random-walk problem of statistics: what is the probability of a person ending a random walk a radius a_T from the starting point at an angle φ_T from a reference axis if each step is of the same magnitude a_k and in a random direction φ_k? The answer was first obtained by Lord Rayleigh [12] in the very connection of the superposition of waves.

The probability $P(a_T, \varphi_T)$ that a set of points within a just resolvable area will yield a resultant amplitude a_T and phase φ_T is given by the Rayleigh distribution

$$P(a_T, \varphi_T) = \frac{1}{\pi N} \exp\left(-a_T^2/N\right). \tag{8.13}$$

Since we are interested in finding the resultant irradiance, a more pertinent probability function is $P(a_T)$, which is the probability of finding a resultant amplitude a_T regardless of the phase. Such a probability is obtained by integrating Eq. 8.13 over the circular annulus between a_T and da_T, giving [13]

$$P(a_T) = \frac{2a_T}{N} \exp\left(-a_T^2/N\right). \tag{8.14}$$

This distribution can now be used to calculate the average value of any function of a_T. In particular

$$\bar{H} = \overline{a_T^2} = \int_0^\infty a_T^2 P(a_T) \, d\varphi_T = \frac{2}{N} \int_0^\infty a_T^3 \exp\left(-a_T^2/N\right) da_T = N. \tag{8.15}$$

Also,

$$\overline{H^2} = \overline{a_T^4} = \int_0^\infty a_T^4 P(a_T) \, da_T = \frac{2}{N} \int_0^\infty a_T^5 \exp\left(-a_T^2/N\right) da_T = 2N^2 \tag{8.16}$$

so that

$$\sigma = [\overline{H^2} - \bar{H}^2]^{1/2} = [2N^2 - N^2]^{1/2} = N = \bar{H} \tag{8.17}$$

and we see that the fluctuations are by no means small and are comparable with the average irradiance itself. The RMS deviation from the mean irradiance in a speckle pattern is simply equal to the average irradiance.

The problem of laser speckle can be quite serious, as is evidenced by the degradation of the image shown in Fig. 8.8, but the problem is many times more severe for long-wavelength holography, such as acoustical holography, where the effective wavelength might be of the order of millimeters. When this happens, even relatively large holograms represent apertures of relatively few wavelengths. This makes the size of the just resolvable spots, and hence the size of the speckles, enormously large, so that in some cases the image is barely distinguishable. In optical holography the problem of laser speckle is most severe in trying to store images as very small holograms (of the order of a millimeter or so), and in doing holographic microscopy. Because of the severity of the problem in these cases, there has been a good deal of work attempting to reduce or eliminate laser speckle from holographic images.

The most straightforward method for eliminating the speckle is to eliminate the diffuse object and use one that is specular, such as a transparency that is not backed by a diffusing plate. In this case there are no random scattering centers to produce the random interference pattern in the image plane. The obvious problem with this is that it removes the redundancy that is the main feature of diffuse object holography. By redundancy we mean that for a diffuse object, the light from each point spreads out so as to cover the whole holo-gram. Hence every point of the hologram contains information about every point of the object and the information is said to be redundant. The value of a redundant hologram is that it is essentially immune to defects such as dust, scratches, and blemishes. A further feature, of course, is that the image can be viewed by simple observation, as if through a window without the need for additional optics. Thus eliminating speckle by eliminating the diffusing plate is usually an unsatisfactory solution.

Another speckle reduction scheme that preserves redundancy in the hologram involves making a very low f/number hologram. The resolution in the image formed with such a hologram is very high so that the size of the just resolvable spot on the object is very small. This can result in a speckle size that is too small to be seen in the final image, or at least rendered unobjectionable. This method, however, entails making a relatively large hologram, which in many cases is not consistent with the application at hand.

In principle one can eliminate speckle by the use of special diffusers [14–16]. An example of such a diffuser is a pure phase modulator as proposed by Upatnieks [14]. The phase modulator is placed in contact with the trans-parency, so that if the holographic imagery is perfect, the modulator is not visible in the image. For this method to work effectively, two conditions must be fulfilled: (1) The diffuser should be such that it only varies the phase of the wavefront and cause no appreciable amplitude variation, and (2) the spread function of the imaging system when projected onto the object plane should be small compared with the size of the variations of phase. In effect, the first of these conditions says that the diffuser should be of a type where there are no high gradients in the surface so that there are internal reflections or multiple scatterings. Ground glass is a typical material that would not fulfill these conditions. The type of glass commonly used in front of portraits is much more suitable, because its surface typically contains many small but smooth undulations.

Converting the second condition into terms relating to the aperture plane, it says that the power spectrum of the amplitude transmittance of the diffuser should be smaller than the Fourier transform of the amplitude spread function; that is, the light distribution in the far field is contained within the aperture of the imaging system.

Finding precisely the right phase diffuser to fulfill these conditions has been, and still is, the subject of much research.

Another type of special diffuser that is easy to construct and to control is simply a phase grating [15, 16]. A phase grating, rather than a diffuser, is placed behind the object transparency so that several defocused images fall within the hologram aperture. This method has been shown to produce good redundancy and good image quality in certain applications, most notably the Selectavision system of RCA [15].

Still another more or less straightforward method for eliminating the diffuser (and hence the speckle) while maintaining redundancy is to form a Fourier transform hologram of a specularly illuminated object transparency. The hologram is redundant in this case because each point of the object is represented by a plane wave propagating at a given angle to the hologram. This plane wave (which interferes with the reference wave) is recorded over the whole area of the hologram, yielding the redundancy. The problem with this method is that the Fourier transform of most transparencies is highly non-uniform, especially at the center where there is usually a very intense hot spot representing the DC component. The dynamic range of the recording material is usually exceeded so that the DC or large-area information is either not recorded at all, or recorded with a diffraction efficiency so different from the rest that the tone reproduction in the image is very poor. The usual way around this trades redundancy for more uniform illumination: the recording plane is simply moved slightly away from the exact Fourier transform plane. An amount equal to 5–10% of the focal length of the transforming lens is usually sufficient.

Laser speckle can be reduced by an averaging technique whereby several holograms of the same object but different diffuser are recorded. If an optical system can be devised such that all of the images are incoherently superposed (such as by successive printing), the independent speckle patterns will tend to average out. However, to obtain an improvement in signal-to-noise ratio by a factor n, a number n^2 holograms must be superposed, thus requiring n^2 as much area.

The final method for eliminating or reducing speckle is one that trades resolution in the image for speckle reduction. This is done by illuminating the hologram with a nonlaser source, such as a filtered arc lamp. In this case the source size and line width are such that the coherence is not great enough to form the speckle pattern. The amount of resolution lost depends on both the source size and bandwidth, as discussed in the next chapter.

This list of methods for reducing laser speckle in holography is not by any means complete. Many workers have attacked this problem and many schemes have been devised to combat this problem. More detailed descriptions and many variations of the above mentioned methods can be found in the literature. None have been completely successful, so that work in this area is continuing.

REFERENCES

[1] E. N. Leith, *Phot. Sci. Eng.*, **6**, 75 (1962).

[2] C. W. Helstrom, *J. Opt. Soc. Am.*, **56**, 433 (1966).

[3] C. B. Burckhardt, *Appl. Opt.*, **6**, 1359 (1967).

[4] G. B. Brandt, *Appl. Opt.*, **9**, 1424 (1970).

[5] K. Biedermann, *Optik*, **31**, 1 (1970).

[6] D. H. R. Vilkomerson, *Appl. Opt.*, **9**, 2080 (1970).

[7] H. M. Smith, *Appl. Opt.*, **11**, 26 (1972).

[8] J. W. Goodman, *J. Opt. Soc. Am.*, **57**, 493 (1967).

[9] A. Kozma, *J. Opt. Soc. Am.*, **58**, 463 (1968).

[10] W. H. Lee and M. O. Green, *J. Opt. Soc. Am.*, **61**, 402 (1971).

[11] R. J. Collier, C. B. Burckhardt, and L. H. Lin, *Optical Holography*, Academic Press, New York, 1971, p. 341.

[12] Lord Rayleigh, *Scientific Papers*, Vol. 1, Cambridge University Press, London, 1899–1920, p. 491.

[13] J. M. Stone, *Radiation and Optics*, McGraw-Hill, New York, 1963, p. 150.

[14] J. Upatnieks, *Appl. Opt.*, **6**, 1905 (1967).

[15] H. J. Gerritsen, W. J. Hannan, and E. G. Ramberg, *Appl. Opt.*, **7**, 2301 (1968).

[16] E. N. Leith and J. Upatnieks, *Appl. Opt.*, **7**, 2085 (1968).

9 Holographic Image Resolution

9.0 INTRODUCTION

In this chapter we examine several of the factors that can limit the resolution in a holographic image, namely, the size and bandwidth of the source (or sources) providing the reference and illuminating beams, the resolution capabilities of the recording medium, and also the geometrical aberrations that can arise. Each of these factors is treated separately, as if it alone were degrading the image, although in actual practice, of course, they act simultaneously. The ultimate, diffraction-limited resolution is derived as a natural consequence of the treatment in Section 9.1. No actual numbers or statements of the realizable resolution are given in Section 9.4, since the actual final image resolution is such a complicated function of the aberrations of the system.

9.1 REFERENCE AND ILLUMINATING BEAM SOURCE SIZE

In all of the analyses presented so far, we have assumed that the reference beam originated at a point source, resulting in a strictly plane or spherical reference beam. The same assumption has been made regarding the illuminating beam. In any real case, however, one never has a true mathematical point source; there is always some finite size associated with the reference and illuminating beam sources. Also, in many cases, the two sources may be physically different, such as when a change in wavelength or a change in divergence is desired. In this section we will determine the effect of nonzero source size on the holographic image. In line with our usual procedure, we treat the problem in two dimensions only, since consideration of the three-dimensional problem only adds to the length of the equations and not to the insight.

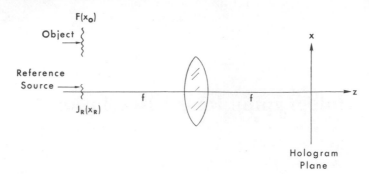

Fig. 9.1 A Fourier transform hologram system. The reference beam is derived from a nonpoint source described by $J_R(x_R)$.

Consider the Fourier transform system of Fig. 9.1. We assume that the lens is large enough so that we can neglect the effects of vignetting. The complex amplitude distribution across the object is represented by $F(x_o)$ and across the reference source by $J_R(x_R)$. Both the object and the reference source are in the same plane, a distance f from the lens of focal length f. The hologram plane is the x-plane and is located a distance f behind the lens. The object field distribution in this plane is (see Eq. A.18)

$$O(x) = \left(\frac{-i}{\lambda f}\right)^{1/2} \int_{-\infty}^{\infty} F(x_o) e^{-i(k/f)xx_o} \, dx_o. \tag{9.1}$$

Similarly, the reference field at the hologram is given by

$$R(x) = \left(\frac{-i}{\lambda f}\right)^{1/2} \int_{-\infty}^{\infty} J_R(x_R) e^{-i(k/f)xx_R} \, dx_R. \tag{9.2}$$

The exposure term leading to the primary image wave is OR^*, and we will consider this term alone. Assuming simple equality between exposure and final amplitude transmittance, illuminating the hologram with a wave $C(x)$ yields a transmitted wave

$$\psi(x) = C(x)O(x)R^*(x). \tag{9.3}$$

We write the illuminating wave in a manner similar to Eq. 5.2 for the reference wave:

$$C(x) = \left(\frac{-i}{\lambda f}\right)^{1/2} \int_{-\infty}^{\infty} J_c(x_c) e^{-i(k/f)xx_c} \, dx_c, \tag{9.4}$$

where $J_c(x_c)$ is the field distribution across the source of the illuminating wave. Substituting Eqs. 9.1, 9.2, and 9.4 into 9.3 leads to an expression for the

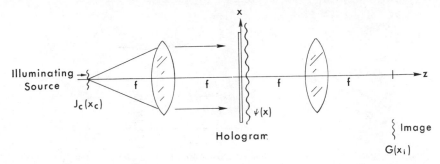

Fig. 9.2 Forming the image with the hologram recorded as in Fig. 9.1. The illuminating beam is derived from a nonpoint source described by $J_c(x_c)$.

transmitted wave:

$$\psi(x) = \frac{1}{\lambda f}\left(\frac{-i}{\lambda f}\right)^{1/2} \int\!\!\!\int\!\!\!\int_{-\infty}^{\infty} J_c(x_c) J_R^*(x_R) F(x_o) e^{-i(k/f)x(x_c - x_R + x_o)}\, dx_c\, dx_R\, dx_o. \quad (9.5)$$

To form an image with this transmitted wave, we use a lens of focal length f as shown in Fig. 9.2. The resulting image $G(x_i)$ is formed in the plane $z = 2f$ and is given by

$$G(x_i) = \left(\frac{-i}{\lambda f}\right)^{1/2} \int_{-\infty}^{\infty} \psi(x) e^{-i(k/f)x x_i}\, dx. \quad (9.6)$$

We have written the limits on this integral as $\pm\infty$ even though the range of x is limited by the finite dimensions of the hologram. The infinite limits will represent a good approximation as long as the hologram size is large enough so that the effects of diffraction are negligible compared to the effects due to the extent of the sources J_R and J_c. We compute these effects separately, beginning with those caused by the finite size of the sources in the absence of diffraction. Later we will return to Eq. 9.6 and determine the effects of diffraction.

We can best determine these effects by finding the *line spread function* for the holographic system, which is just the irradiance distribution across the image of an infinitely thin line object. This gives us the best estimate of the resolution capabilities of the system. We first find the amplitude distribution in the image of a line. Substituting Eq. 9.5 into 9.6 gives

$$G(x_i) = \frac{-1}{\lambda^2 f^2} \int\!\!\!\int\!\!\!\int\!\!\!\int_{-\infty}^{\infty} J_c(x_c) J_R^*(x_R) F(x_o) e^{-i(k/f)x(x_c - x_R + x_o + x_i)}\, dx_c\, dx_R\, dx_o\, dx.$$

$$(9.7)$$

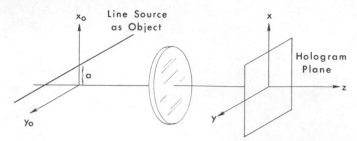

Fig. 9.3 Recording a Fourier transform hologram of an infinitely narrow line object.

Writing a line object located a distance a above the axis (see Fig. 9.3) as

$$F(x_o) = \delta(x_o - a),\qquad(9.8)$$

where $\delta(x)$ is the Dirac delta function, the amplitude distribution in the image becomes

$$G_l(x_i) = \frac{-1}{\lambda^2 f^2} \int\!\!\!\int\!\!\!\int\!\!\!\int_{-\infty}^{\infty} J_c(x_c) J_R^*(x_R)\, \delta(x_o - a) e^{-i(k/f)x(x_c - x_R + x_o + x_i)}\, dx_c\, dx_R\, dx_o\, dx$$

$$= \frac{-1}{\lambda^2 f^2} \int\!\!\!\int\!\!\!\int_{-\infty}^{\infty} J_c(x_c) J_R^*(x_R) e^{-i(k/f)x(x_c - x_R + a + x_i)}\, dx_c\, dx_R\, dx,\qquad(9.9)$$

where we have written $G_l(x_i)$ to denote that this is the image of a line object. Using the relation

$$\delta(x) = \frac{1}{2\pi} \int_{-\infty}^{\infty} e^{-ikx}\, dk,\qquad(9.10)$$

we see that

$$\int_{-\infty}^{\infty} e^{-i(k/f)x(x_c - x_R + a + x_i)}\, dx = 2\pi\, \delta\!\left[\frac{k}{f}(x_c - x_R + a + x_i)\right]$$

$$= \lambda f\, \delta(x_c - x_R + a + x_i)\qquad(9.11)$$

so that

$$G_l(x_i) = \frac{-1}{\lambda f} \int\!\!\!\int_{-\infty}^{\infty} J_c(x_c) J_R^*(x_R)\, \delta(x_c - x_R + a + x_i)\, dx_c\, dx_R.\qquad(9.12)$$

Integrating this over x_c gives

$$G_l(x_i) = \frac{-1}{\lambda f} \int_{-\infty}^{\infty} J_c(x_R - x_i - a) J_R^*(x_R)\, dx_R.\qquad(9.13)$$

Hence we see that the amplitude spread function of the holographic system is given by a correlation of the reference and illuminating beam source distributions, J_c and J_R. The line spread function is just $|G_i(x_i)|^2$. A full three-dimensional analysis would give rise to a two-dimensional correlation integral in place of Eq. 9.13, and would lead to the point spread function, or impulse response, of the holographic system. The correlation integral indicates that, in general, if the reference source has a width A and the illuminating source a width B, then the line spread function (in amplitude) of the system will have a width of the order of $A + B$, the maximum possible resolution of this system will be approximately $1/(A + B)$ resolvable lines per unit distance.

If the illuminating beam is derived from a point source (or in reality a very small source), but the reference source has a finite extent, then

$$J_c(x_c) = \delta(x_c) \tag{9.14}$$

and

$$G_i(x_i) = \frac{-1}{\lambda f} \int_{-\infty}^{\infty} \delta(x_R - x_i - a) J_R^*(x_R)\, dx_R$$

$$= \frac{-1}{\lambda f} J_R^*(x_i + a) \tag{9.15}$$

so that an image of a line at a distance a above the axis is the conjugate of the reference source distribution centered a distance a below the axis. Hence we can say that for illumination of the hologram with an ideal plane or spherical wave, the minimum resolvable line image can be no less than the extent of the reference source. Conversely, we can also say that if the reference wave is ideally plane or spherical, but the illuminating wave is derived from a source other than a point source, then the line object will image as the illuminating source distribution, $J_c(x_c)$.

We next wish to determine the line spread function for diffraction-limited imagery. We begin by noting that in the ideal case both J_c and J_R are sufficiently small that we can write

$$J_R(x_R) = \delta(x_R)$$

and

$$J_c(x_c) = \delta(x_c). \tag{9.16}$$

With this approximation Eq. 9.7 becomes

$$G(x_i) = -\frac{1}{\lambda^2 f^2} \int\!\!\!\int\!\!\!\int\!\!\!\int_{-\infty}^{\infty} \delta(x_c)\, \delta(x_R) F(x_o) e^{-i(k/f)x(x_c - x_R + x_o + x_i)}\, dx_c\, dx_R\, dx_o\, dx$$

$$= -\frac{1}{\lambda^2 f^2} \int\!\!\!\int_{-\infty}^{\infty} F(x_o) e^{-i(k/f)x(x_o + x_i)}\, dx_o\, dx. \tag{9.17}$$

As long as the limits on the x-integration are $\pm\infty$, we may use the definition (9.10) and write

$$
\begin{aligned}
G(x_i) &= -\frac{2\pi}{\lambda^2 f^2} \int_{-\infty}^{\infty} F(x_o)\, \delta\left[\frac{k}{f}(x_o + x_i)\right] dx_o \\
&= -\frac{2\pi}{\lambda^2 f^2}\frac{f}{k} \int_{-\infty}^{\infty} F(x_o)\, \delta(x_o + x_i)\, dx_o \\
&= -\frac{1}{\lambda f} F(-x_i)
\end{aligned}
\tag{9.18}
$$

and the imagery is perfect. To account for the effects of diffraction, we note that in any real case the hologram plane only extends from $-H$ to H so that Eq. 9.17 should be written

$$
\begin{aligned}
G(x_i) &= \frac{-1}{\lambda^2 f^2} \int_{-\infty}^{\infty} F(x_o)\left[\int_{-H}^{H} e^{-i(k/f)x(x_o + x_i)}\, dx\right] dx_o \\
&= \frac{-2H}{\lambda^2 f^2} \int_{-\infty}^{\infty} F(x_o)\, \mathrm{sinc}\left[\frac{kH}{f}(x_o + x_i)\right] dx_o.
\end{aligned}
\tag{9.19}
$$

For an infinitely narrow line object located at $(x_o, z_o) = (a, -2f)$, we have

$$
F(x_o) = \delta(x_o - a)
\tag{9.20}
$$

and the diffraction-limited, line amplitude spread function becomes

$$
\begin{aligned}
G_l(x_i) &= \frac{-2H}{\lambda^2 f^2} \int_{-\infty}^{\infty} \delta(x_o - a)\, \mathrm{sinc}\left[\frac{kH}{f}(x_o + x_i)\right] dx_o \\
&= \frac{-2H}{\lambda^2 f^2}\, \mathrm{sinc}\left[\frac{kH}{f}(x_i + a)\right].
\end{aligned}
\tag{9.21}
$$

The line spread function of the system is now just

$$
|G_l(x_i)|^2 = \frac{4H^2}{\lambda^4 f^4}\, \mathrm{sinc}^2\left[\frac{kH}{f}(x_i + a)\right].
\tag{9.22}
$$

If a three-dimensional analysis had been made, the image of a point object located at $x_o = a$, $y_o = b$ would be given by

$$
G_p(x_i, y_i) = \frac{-4H^2}{\lambda^4 f^4}\, \mathrm{sinc}\left[\frac{kH}{f}(x_i + a)\right] \mathrm{sinc}\left[\frac{kH}{f}(y_i + b)\right]
\tag{9.23}
$$

and the point spread function for the system would be just $|G_p(x_i, y_i)|^2$. The line spread function (9.22) is centered about the point $x_i = -a$. The width of the line image can be taken as the distance between the zeros on either side of the central maximum of the sinc function. These occur for the argument equal to $\pm\pi$:

$$\frac{kH}{f}(x_i + a) = \pm\pi. \tag{9.24}$$

The width of the line image is thus $f\lambda/H$. Hence the number of resolvable lines per unit distance will be of the order of $H/f\lambda$.

The most common situation is where both the reference and illuminating beams are derived from the same source, so that $J_c = J_R$. In this case Eq. 9.13 becomes

$$G_i(x_i) = \frac{-1}{\lambda f}\int_{-\infty}^{\infty} J_R(x_R - x_i - a)J_R^*(x_R)\,dx_R. \tag{9.25}$$

Some interesting conclusions may be drawn from this relationship, but first we write Eq. 9.2 as a Fourier transform. Since Eq. 9.2 is in the same form as Eq. A.18 of the Appendix (except for the constant phase factor e^{2ikf}), we can see that the desired form is

$$V(\nu) = \int_{-\infty}^{\infty} J_R(x_R)e^{-2\pi i\nu x_R}\,dx_R \tag{9.26}$$

where

$$V(\nu) = \left(\frac{-i}{\lambda f}\right)^{-1/2} \cdot R(\lambda f\nu) \tag{9.27}$$

and

$$\nu = \frac{x}{\lambda f}. \tag{9.28}$$

By Fourier inversion of Eq. 9.26 we can write

$$J_R(x_R) = \int_{-\infty}^{\infty} V(\nu)e^{2\pi i\nu x_R}\,d\nu \tag{9.29}$$

so that from Eq. 9.25

$$-\lambda f G_i(x_i) = \int\!\!\!\int_{-\infty}^{\infty} V(\nu)e^{2\pi i\nu(x_R - x_i - a)}\,d\nu\,dx_R \int_{-\infty}^{\infty} V^*(\nu')e^{-2\pi i\nu' x_R}\,d\nu'. \tag{9.30}$$

Interchanging the order of integration and combining the exponentials, we get

$$-\lambda f G_i(x_i) = \int\!\!\!\int\!\!\!\int_{-\infty}^{\infty} V(\nu)V^*(\nu')e^{-2\pi i\nu(x_i + a)}e^{-2\pi i x_R(\nu' - \nu)}\,d\nu\,d\nu'\,dx_R$$

$$= \int\!\!\!\int_{-\infty}^{\infty} V(\nu)V^*(\nu')e^{-2\pi i\nu(x_i + a)}\,\delta(\nu' - \nu)\,d\nu\,d\nu'$$

$$= \int_{-\infty}^{\infty} |V(\nu)|^2\, e^{-2\pi i\nu(x_i + a)}\,d\nu. \tag{9.31}$$

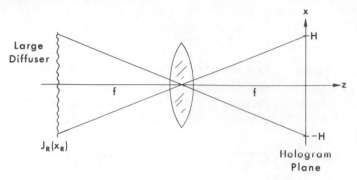

Fig. 9.4 The use of a large diffuser as the reference source.

Recall that $G_l(x_i)$ is the line amplitude spread function of the holographic system and that $V(\nu)$ is closely related to the light distribution at the hologram plane due to the reference beam alone, and that $V(\nu)$ is the Fourier transform of $J_R(x_R)$, which is the field distribution across the reference source. Since G_l is essentially an autocorrelation function (Eq. 9.25), it is not surprising that it can be written as the Fourier transform of the energy spectrum, $|V(\nu)|^2$. It is interesting to note the form of the line amplitude spread function when $V(\nu)$ is band limited. Suppose that the hologram extends from $-H$ to H so that the range of spatial frequencies that can be recorded is only $-\nu_m$ to ν_m where

$$\nu_m = \frac{H}{\lambda f}. \tag{9.32}$$

Now suppose that the reference source is a diffuser sufficiently large that the complete frequency range $2\nu_m$ is covered (Fig. 9.4). If the diffuser is perfect, that is, completely random, $J_R(x_R)$ is in the form of a white noise so that

$$|V(\nu)|^2 = V_0, \qquad -\nu_m \le \nu \le \nu_m$$
$$= 0 \qquad \text{elsewhere} \tag{9.33}$$

In this case Eq. 5.31 becomes

$$-\lambda f G_l(x_i) = \int_{-\nu_m}^{\nu_m} V_0 e^{-2\pi i \nu(x_i + a)} \, d\nu \tag{9.34}$$

so that

$$G_l(x_i) = -V_0 \frac{2\nu_m}{\lambda f} \operatorname{sinc} [2\pi \nu_m(x_i + a)]. \tag{9.35}$$

Using Eq. 9.32 we find

$$G_l(x_i) = -V_0 \frac{2H}{\lambda^2 f^2} \operatorname{sinc} \left[\frac{kH}{f}(x_i + a)\right] \tag{9.36}$$

which, except for the constant V_o, is exactly the same as Eq. 9.21. This means that the image resolution should be the same whether one uses a point reference source or a perfectly diffuse one. The assumption that the diffuser be perfect is impossible, however, since this would require that both the diffuser and lens of Fig. 9.4 be infinitely large. This is because any finite sized diffuser leads to a speckle pattern in the hologram plane, which means that $|V(v)|^2$ is not constant. This speckle is, of course, just the random interference pattern that results from the interference of all pairs of points of the diffuser. In actual practice, however, the resolution that can be achieved with a diffuse reference beam is about equal to that which can be obtained with a small reference source. The point spread function for such a system is essentially the diffraction-limited point spread function, surrounded by uniform flare.

There is an important reciprocal relationship between source size and image resolution that should be mentioned here; it may be stated simply as follows. If the correlation function (Eq. 9.25) is very sharply peaked, indicating either a very small or very large and diffuse reference source distribution, the size of the hologram that can be filled with light is very large, resulting in only a small amount of light spread caused by diffraction. Hence the resolution will be high. On the other hand, if the correlation function (9.25) is very broad, indicating a large source, only a small hologram area can be filled, resulting in poor resolution because of diffraction.

As an example of this effect, consider the common situation where a microscope objective is used to diverge the reference beam so that it will fill the hologram. The spot at the focus of the objective then serves as the source for the reference beam. For the TEMoo laser mode, the spot size at the focus of the objective is given by

$$w_o = \frac{\lambda}{\pi \text{ N.A.}} \tag{9.37}$$

where N.A. is the numerical aperture of the objective (see Eq. 10.29 and following) and w_o is the radius of the spot measured to the $1/e$ amplitude points. But w_o is related to the hologram size $2H$ and the focal length f of the lens in Fig. 9.1 through

$$\text{N.A.} = \frac{H}{f} = \frac{\lambda}{\pi w_o} \tag{9.38}$$

so that

$$w_o = \frac{\lambda f}{\pi H}. \tag{9.39}$$

The width of the correlation peak of Eq. 9.25 is thus roughly

$$4w_o = \frac{4\lambda f}{\pi H}. \tag{9.40}$$

The half-width of a line image for a diffraction-limited hologram is $f\lambda/H$ (compare Eq. 9.24). Comparison of this result with Eq. 9.40 indicates that the two resolution limits are essentially equal. The conclusion to be drawn from this is that nothing will be gained in terms of image resolution by using a higher N.A. objective than is necessary to just fill the hologram with light. Overfilling the hologram by use of a high N.A. objective will not increase the resolution, even though the reference source size will be small.

9.2 REFERENCE AND ILLUMINATING BEAM BANDWIDTH

It is well known that high-quality holographic images require a narrow-band light source, both for recording and illuminating the hologram. An ideal light source is a point source of zero size and zero bandwidth. Of course, actual sources used in practice are only approximations to this ideal. Sharpness in an image is always limited by diffraction, so the best possible light source need be no smaller than the size dictated by diffraction effects, as discussed above. Further, the bandwidth need not be zero, but only small enough so that any spreading of light in the image caused by finite bandwidth be small compared to the spreading caused by diffraction.

We can obtain an estimate of the resolution-bandwidth relationship by considering a simple diffraction grating model of a hologram. Suppose two plane waves are incident on the hologram plane at some angle α_R as shown in Fig. 9.5. The resulting interferogram constitutes a Fourier transform hologram of a point object, and the fringes recorded on the hologram vary spatially in transmittance in an approximately sinusoidal manner. The spatial frequency of these fringes is given by

$$v_s = \frac{\sin \alpha_R}{\lambda} \qquad (9.41)$$

where λ is the recording wavelength. If the recording light has a finite

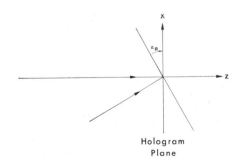

Fig. 9.5   A simple plane-wave hologram.

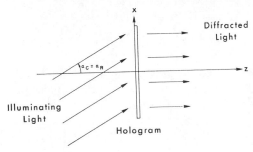

Fig. 9.6 Diffraction of the primary wave from the grating (hologram) recorded as in Fig. 9.5. The diffracted wave makes an angle $\alpha_i = 0$ with the z-axis if the reference and illuminating beams are monochromatic.

bandwidth $\Delta\lambda_R$, then

$$\lambda_R - \frac{\Delta\lambda_R}{2} \leq \lambda \leq \lambda_R + \frac{\Delta\lambda_R}{2} \tag{9.42}$$

where λ_R is the central wavelength of the recording light. This wavelength spread gives rise to a spread in recorded spatial frequency given by

$$\Delta\nu_s = -\frac{\sin\alpha_R}{\lambda_R^2}\Delta\lambda_R. \tag{9.43}$$

To reconstruct, we illuminate the hologram with a plane wave incident at the angle $\alpha_c = \alpha_R$ and wavelength λ (Fig. 9.6) such that

$$\lambda_c - \frac{\Delta\lambda_c}{2} \leq \lambda \leq \lambda_c + \frac{\Delta\lambda_s}{2} \tag{9.44}$$

where $\Delta\lambda_c$ is the bandwidth of the illuminating beam and λ_c the central wavelength. The angle of diffraction α_i of the first-order (primary) wave is given by

$$\sin\alpha_i = \lambda\nu_s - \sin\alpha_c, \tag{9.45}$$

which is equal to zero for monochromatic light, as shown in Fig. 9.6. However, because of the spread in wavelengths of the illuminating and recording beams, there is a spread in diffraction angles

$$\Delta(\sin\alpha_i) = (\sin\alpha_i)_{\text{max}} - (\sin\alpha_i)_{\text{min}}, \tag{9.46}$$

where

$$(\sin\alpha_i)_{\text{max}} = \lambda_{\text{max}}\nu_{s_{\text{max}}} = \left(\lambda_c + \frac{\Delta\lambda_c}{2}\right)\left[\frac{\sin\alpha_R}{\lambda_R} + \frac{\sin\alpha_R}{2\lambda_R^2}\Delta\lambda_R\right] \tag{9.47}$$

and

$$(\sin\alpha_i)_{\text{min}} = \lambda_{\text{min}}\nu_{s_{\text{min}}} = \left(\lambda_c - \frac{\Delta\lambda_c}{2}\right)\left[\frac{\sin\alpha_R}{\lambda_R} - \frac{\sin\alpha_R}{2\lambda_R^2}\Delta\lambda_R\right] \tag{9.48}$$

so

$$\Delta(\sin \alpha_i) = \frac{\lambda_c}{\lambda_R}\left[\frac{\Delta\lambda_R}{\lambda_R} + \frac{\Delta\lambda_c}{\lambda_c}\right] \sin \alpha_R. \tag{9.49}$$

We see that the finite bandwidths of both recording and illuminating beams led to an angular spread in the reconstructed object wave. In terms of the optical frequencies involved, this spread is given by

$$\Delta(\sin \alpha_i) = \frac{\nu_R}{\nu_c}\left[\frac{\Delta\nu_c}{\nu_c} + \frac{\Delta\nu_R}{\nu_R}\right] \sin \alpha_R. \tag{9.50}$$

A common situation is that the reference and illuminating beams are identical. In this case $\nu_R = \nu_c$ and $\Delta\nu_R = \Delta\nu_c$ and so

$$\Delta(\sin \alpha_i) = 2\frac{\Delta\nu_R}{\nu_R} \sin \alpha_R. \tag{9.51}$$

For a multimode He-Ne gas laser, the effective $\Delta\nu_R$ is of the order of 1.5×10^9 Hz. This leads to

$$\Delta(\sin \alpha_i) \approx 6 \times 10^{-6} \sin \alpha_R \approx \Delta\alpha_i. \tag{9.52}$$

This is, of course, very small, especially in view of the fact that the angular spread due to diffraction, $\Delta\theta_D$, is given by

$$\Delta\theta_D \approx \frac{\lambda}{2H} = \frac{c}{2H\nu_c} \tag{9.53}$$

where $2H$ is the hologram aperture. For the He-Ne gas laser, the angular spread due to finite bandwidth becomes equal to the spread due to diffraction for a hologram aperture

$$2H \simeq 20 \text{ cm}. \tag{9.54}$$

For the argon laser, the situation is slightly worse because of a factor of about 3 in $\Delta\nu_R$. For this laser, the two angular spreads are equal for an aperture of

$$2H \simeq 5 \text{ cm}. \tag{9.55}$$

The spread of directions $\Delta\alpha_i$ of the diffracted wave means that this wave will focus to a line width $f\Delta\alpha_i$ (Fig. 9.7). This will be the width of the line spread function of the system.

Although we have confined the discussion to a simple diffraction grating hologram, the results are easily extended to more complex object distributions. For a Fourier transform hologram of an extended object, each object point records as a single spatial frequency at the hologram plane, so that the above analysis can be applied directly. For a Fresnel hologram, we may use the above results if we consider the hologram as a superposition of very many

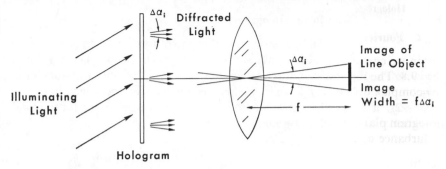

Fig. 9.7 The line spread function for a hologram system in which the hologram is recorded and illuminated with light of nonzero bandwidth. This gives rise to a spread in diffraction angles $\Delta\alpha_i$ and a line spread function of width $f\Delta\alpha_i$.

simple gratings. Each object point is now recorded as a range of spatial frequencies at the hologram plane. The range of spatial frequencies given by Eq. 9.43 can now be regarded as the spread in recorded frequency at each small element of area of the hologram. There is now a different $\Delta\alpha_i$ at each point of the hologram, but the resulting spread function is substantially the same as for a Fourier hologram for most recording configurations. The line spread function for a Fresnel hologram has an approximate width given by $\overline{\Delta\alpha_i} \cdot z_o$, where $\overline{\Delta\alpha_i}$ is the average spread in the diffracted beam and z_o is the image distance.

9.3 EFFECT OF THE RECORDING MEDIUM [1, 2]

9.3.1 Introduction

Up to this point, we have tacitly assumed that the recording medium is capable of resolving all of the spatial frequencies of interest, except possibly for a cutoff caused by the finite extent of the hologram. This assumption, of course, represents an idealized situation, since every recording medium has some upper limit on the spatial frequencies it is capable of recording. One of the most useful measures of the ability of a recording medium to resolve fine detail is its modulation transfer function (MTF). This is simply the ratio of output to input sine wave modulations, expressed as a function of spatial frequency. If it is defined in one dimension, it is the Fourier transform of the line spread function. The purpose of this section, then, is to determine the effect of the MTF of the recording medium on the resolution in the holographic image.

9.3.2 Fourier Transform Holograms

Again we consider a basic Fourier transform arrangement such as shown in Fig. 9.8. The reference beam makes an angle α_R with the axis as shown, and the complex object distribution is denoted by $F(x_o)$. The object is situated at $z_o = -2f$, a distance f from a well-corrected lens of focal length f. The hologram plane is a distance f from the lens, so that we may write for the total disturbance at the hologram plane

$$H(x) = R(x) + O(x) = e^{-ikx \sin \alpha_R} + \left(\frac{-i}{\lambda f}\right)^{1/2} \int_{-\infty}^{\infty} F(x_o)e^{-i(k/f)xx_o} \, dx_o. \quad (9.56)$$

The incident exposure is proportional to $|H(x)|^2$, and the effective exposure, because of the limited MTF of the recording medium, will be a convolution of the incident exposure and the spread function of the recording medium. In one dimension, then, the effective exposure is

$$E(x) = \int_{-\infty}^{\infty} S(u) \, |H(x - u)|^2 \, du, \quad (9.57)$$

where $S(u)$ is the line spread function of the medium.* In order to eliminate as many unimportant constants as possible, we will assume that, after processing, the resulting amplitude transmittance of the hologram is simply equal

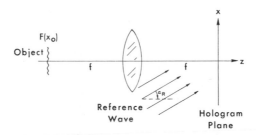

Fig. 9.8 A basic Fourier transform hologram arrangement.

* There is some question here as to whether this convolution can be carried out as indicated or whether one must use an *amplitude spread function* convolved with the light *amplitude* at the hologram. The reason for the question is that the exposing light is both spatially and temporally coherent, which means that one must add amplitudes and then square the sum to arrive at the irradiance. On the other hand, Eq. 9.57 implies that each elementary line in the object images to an irradiance distribution and that the elementary $S(u)$ functions simply add to give the resultant irradiance distribution. Whether one is justified in using 9.57 will have to await future developments; the data taken by van Ligten [1] show a discrepancy with the theory which might be explained along these lines, although Vander Lugt's [3] work shows good agreement with theory.

to the exposure. In this case

$$t(x) = E(x) = \int_{-\infty}^{\infty} S(u)[|R(x-u)|^2 + |O(x-u)|^2$$
$$+ R^*(x-u)O(x-u) + R(x-u)O^*(x-u)] \, du$$
$$\equiv I_0 + I_1 + I_2 \tag{9.58}$$

where

$$I_0 = \int_{-\infty}^{\infty} S(u)[|R(x-u)|^2 + |O(x-u)|^2] \, du \tag{9.59}$$

$$I_1 = \int_{-\infty}^{\infty} S(u)[R^*(x-u)O(x-u)] \, du \tag{9.60}$$

$$I_2 = \int_{-\infty}^{\infty} S(u)[R(x-u)O^*(x-u)] \, du \tag{9.61}$$

Identification of these three components follows easily from earlier discussion: I_0 is the zero-order bias plus flare light, I_1 is the primary image wave, and I_2 is the conjugate image wave. Considering only $I_1(x)$ for the primary image wave, we have

$$I_1(x) = \int_{-\infty}^{\infty} S(u)[R^*(x-u)O(x-u)] \, du$$
$$= \left(\frac{-i}{\lambda f}\right)^{1/2} \int_{-\infty}^{\infty} S(u)e^{ik \sin \alpha_R(x-u)} \int_{-\infty}^{\infty} F(x_o)e^{-i(k/f)x_o(x-u)} dx_o \, du$$
$$= \left(\frac{-i}{\lambda f}\right)^{1/2} e^{ikx \sin \alpha_R} \int_{-\infty}^{\infty} S(u)e^{i(k/f)u(x - f \sin \alpha_R)} du \int_{-\infty}^{\infty} F(x_o)e^{-i(k/f)xx_o} dx_o \tag{9.62}$$

Since the MTF of the recording medium is just the Fourier transform of the line spread function $S(u)$, we have

$$\text{MTF} = \tilde{S}(v) = \int_{-\infty}^{\infty} S(u)e^{-2\pi ivu} \, du, \tag{9.63}$$

where the tilde implies a Fourier transform. Thus we have

$$\tilde{S}\left[\frac{k}{2\pi f}(f \sin \alpha_R - x_o)\right] = \int_{-\infty}^{\infty} S(u)e^{i(k/f)u(x_o - f \sin \alpha_R)} du, \tag{9.64}$$

and 9.62 becomes

$$I_1(x) = \left(\frac{-i}{\lambda f}\right)^{1/2} e^{ikx \sin \alpha_R} \int_{-\infty}^{\infty} \tilde{S}\left[\frac{k}{2\pi f}(f \sin \alpha_R - x_o)\right] F(x_o)e^{-i(k/f)xx_o} dx_o. \tag{9.65}$$

The image is formed in the usual way: a second lens takes the Fourier transform of the transmitted field distribution in the x-plane. If the hologram is illuminated with a wave identical to the reference wave, then the transmitted wave is

$$\psi(x) = e^{-ikx \sin \alpha_R} I_1(x) \tag{9.66}$$

and the image is

$$G(x_i) = \left(\frac{-i}{\lambda f}\right)^{1/2} \int_{-\infty}^{\infty} \psi(x) e^{-i(k/f)xx_i} \, dx. \tag{9.67}$$

Substituting Eq. 9.65 into Eq. 9.66, we obtain

$$G(x_i) = \frac{-i}{\lambda f} \iint_{-\infty}^{\infty} \tilde{S}\left[\frac{k}{2\pi f}(f \sin \alpha_R - x_o)\right] F(x_o) e^{-i(k/f)x(x_i + x_o)} \, dx \, dx_o. \tag{9.68}$$

Using the relation

$$\delta(x) = \frac{1}{2\pi} \int_{-\infty}^{\infty} e^{-ikx} \, dk \tag{9.69}$$

for the Dirac δ-function, the image becomes

$$G(x_i) = \frac{-2\pi i}{\lambda f} \int_{-\infty}^{\infty} \tilde{S}\left[\frac{k}{2\pi f}(f \sin \alpha_R - x_o)\right] F(x_o) \, \delta\left[\frac{k}{f}(x_o + x_i)\right] dx_o$$

$$= -i\tilde{S}\left[\frac{\sin \alpha_R}{\lambda} + \frac{x_i}{\lambda f}\right] F(-x_i). \tag{9.70}$$

Thus we obtain the result that for a Fourier transform hologram, recorded on a medium with limited MTF, the image resolution is not affected, but the field of view is limited.

As a simple example of this effect, consider an MTF given by

$$\tilde{S}(\nu) = 1, \quad -\nu_c \le \nu \le \nu_c$$
$$= 0 \quad \text{otherwise.} \tag{9.71}$$

We wish to make a hologram of a sinusoidly varying object*

$$F(X_o) = 1 + M \cos 2\pi \nu_o X_o \tag{9.72}$$

where M is the modulation of the pattern and ν_o the spatial frequency. From Eq. 5.70 we see that the image is identical to the object in the range

$$-\lambda f \nu_c - f \sin \alpha_R \le X_i \le \lambda f \nu_c - f \sin \alpha_R. \tag{9.73}$$

---

* Strictly speaking, there should be a random phase factor multiplying this expression, implying diffuse illumination. The reason for this is that if this object were specularly illuminated, $O(x)$ would consist only of three very small points of light and no hologram could be formed.

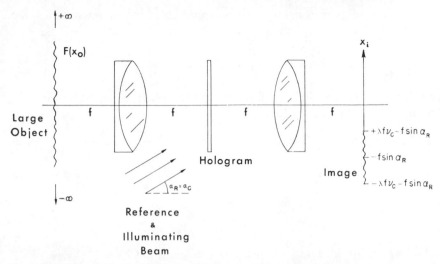

Fig. 9.9 Field-of-view limitation caused by recording a hologram of a large object on a recording medium that can only record spatial frequencies up to ν_c. Although the object extends between $\pm\infty$, the image in the x_i-plane extends only from $\lambda f\nu_c - f\sin\alpha_R$ to $-\lambda f\nu_c - f\sin\alpha_R$.

We see that the image is independent of the object frequency ν_o, but that it is limited in extent in an asymmetrical manner. The situation is illustrated in Fig. 9.9. The object is assumed to extend to $\pm\infty$ in the x_o plane, but only a finite portion is imaged in the x_i plane. The diameter of both lenses and the hologram are assumed to be large enough so that aperture effects can be neglected.

9.3.3 Fresnel Holograms

In this case we consider the arrangement illustrated in Fig. 9.10. A plane object is defined by the complex amplitude distrubution $F(x_o)$ and is located in the x_o plane a distance $-z_o$ from the hologram. A plane reference wave is used, and this wave makes an angle α_R with the axis. We assume that z_o is large

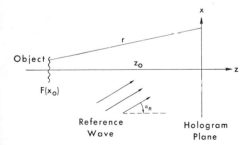

Fig. 9.10 The notation used to describe the recording of a Fresnel hologram.

enough that we may use the Fresnel approximations and write, for the amplitude distribution at the hologram caused by the object,

$$O(x) = \left(\frac{-i}{\lambda}\right)^{1/2} \int_{-\infty}^{\infty} F(x_o) \frac{e^{ikr}}{r} \, dx_o. \tag{9.74}$$

The distance between a point x on the hologram and a point x_o on the object is denoted by r and can be written as

$$r = \pm[(x - x_o)^2 + z_0^2]^{1/2} \approx \pm z_o \pm \frac{(x - x_o)^2}{2z_o} \tag{9.75}$$

Now since $|z_o| = -z_o$ for our coordinate system, we must choose the negative signs to retain our phase convention (compare Eq. A.5).

The reference wave is written as usual

$$R(x) = e^{-ikx \sin \alpha_R} \tag{9.76}$$

and we again consider only the primary image term

$$I_1(x) = \int_{-\infty}^{\infty} S(u)[R^*(x - u)O(x - u)] \, du.$$

Substituting Eqs. 9.74, 9.75, and 9.76, this becomes

$$I_1(x) = \left(\frac{i}{\lambda z_o}\right)^{1/2} \int_{-\infty}^{\infty} S(u)e^{ik(x-u)\sin \alpha_R}$$
$$\times \left[\int_{-\infty}^{\infty} F(x_o)e^{-ikz_0} \exp\left(-ik\frac{(x-u-x_o)^2}{2z_o}\right) dx_o\right] du \tag{9.77}$$

where we have taken $1/r \approx 1/-z_o$. Now define

$$\tilde{F}(v) = \int_{-\infty}^{\infty} F(x_o)e^{-2\pi i v x_o} \, dx_o \tag{9.78}$$

so that

$$F(x_o) = \int_{-\infty}^{\infty} \tilde{F}(v)e^{2\pi i v x_o} \, dv. \tag{9.79}$$

We then have

$$I_1(x) = K \int_{-\infty}^{\infty} S(u)e^{ik(x-u)\sin \alpha_R}$$
$$\times \iint_{-\infty}^{\infty} \tilde{F}(v)e^{2\pi i v x_o} \exp\left(-ik\frac{(x-u-x_o)^2}{2z_o}\right) dv \, dx_o \, du, \tag{9.80}$$

where we have defined

$$K = \left(\frac{i}{\lambda z_o}\right)^{1/2} e^{-ikz_o}. \tag{9.81}$$

Now

$$\int_{-\infty}^{\infty} e^{-i(k/2z_0)(x-u-x_0)^2} e^{2\pi i v x_0}\, dx_o = \left(\frac{\lambda z_0}{i}\right)^{1/2} e^{2\pi i v(x-u)} e^{2i(z_0/k)\pi^2 v^2} \qquad (9.82)$$

so

$$I_1(x) = e^{-ikz_0}\int_{-\infty}^{\infty} S(u) e^{ik(x-u)\sin \alpha_R}\left[\int_{-\infty}^{\infty} \tilde{F}(v) e^{2\pi i v(x-u)} e^{2i(z_0/k)\pi^2 v^2}\, dv\right] du$$

$$= e^{-ikz_0} e^{ikx \sin \alpha_R}\int_{-\infty}^{\infty} S(u) e^{-iku(\sin \alpha_R + 2\pi v/k)}\, du \int_{-\infty}^{\infty} \tilde{F}(v) e^{2\pi i v x} e^{2i(z_0/k)\pi^2 v^2}\, dv$$

$$(9.83)$$

Now define

$$\tilde{S}(v) = \int_{-\infty}^{\infty} S(u) e^{-2\pi i v u}\, du \equiv \text{MTF} \qquad (9.84)$$

so that

$$\tilde{S}\left[v + \frac{k \sin \alpha_R}{2\pi}\right] = \int_{-\infty}^{\infty} S(u) e^{-iku(\sin \alpha_R + 2\pi v/k)}\, du \qquad (9.85)$$

and

$$I_1(x) = e^{-ikz_0} e^{ikx \sin \alpha_R}\int_{-\infty}^{\infty} \tilde{S}\left(v + \frac{\sin \alpha_R}{\lambda}\right) \tilde{F}(v) e^{2\pi i v x} e^{2i(z_0/k)\pi^2 v^2}\, dv. \quad (9.86)$$

To reconstruct, we illuminate the hologram with a wave described by $C(x)$ in the hologram plane. The transmitted field distribution is then

$$\psi(x) = C(x) I_1(x) \qquad (9.87)$$

corresponding, of course, only to the primary image wave. The field in a plane x_i at a distance z_i from the hologram (Fig. 9.11) is given by

$$G(x_i) = \left(\frac{-i}{\lambda}\right)^{1/2}\int_{-\infty}^{\infty} \psi(x)\, \frac{e^{iks}}{s}\, dx \qquad (9.88)$$

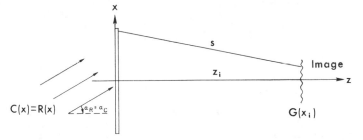

Fig. 9.11 The notation used to describe the reconstruction for the hologram recorded as in Fig. 9.10. The illuminating wave is identical to the reference wave used in recording.

where s is the distance between x and x_i:

$$s = \pm[(x - x_i)^2 + z_i^2]^{1/2} \approx \pm z_i \pm \frac{(x - x_i)^2}{2z_i}, \qquad (9.89)$$

and this time we use the positive root.

If we now assume that the illuminating wave $C(x)$ is identical to the reference wave (Eq. 9.76), $C(x) = R(x)$ and substitution of Eq. 9.87 into (9.88) yields, with the approximation (9.89),

$$G(x_i) = \left(\frac{-i}{\lambda z_i}\right)^{1/2} \int_{-\infty}^{\infty} R(x)I_1(x)e^{iks} \, dx$$

$$= \left(\frac{-i}{\lambda z_i}\right)^{1/2} e^{-ik(z_o - z_i)}$$

$$\times \iint_{-\infty}^{\infty} \tilde{S}\left(v + \frac{\sin \alpha_R}{\lambda}\right) \tilde{F}(v)e^{2\pi iv x}e^{2i(z_o/k)\pi^2 v^2}\exp\left(ik\frac{(x - x_i)^2}{2z_i}\right) \, dv \, dx.$$

$$(9.90)$$

where we have used $1/s \approx 1/z_o$. But

$$\int_{-\infty}^{\infty} e^{i(k/2z_i)(x - x_i)^2}e^{2\pi iv x} \, dx = (i\lambda z_i)^{1/2}e^{2\pi iv x_i}e^{-2i(z_i/k)\pi^2 v^2}, \qquad (9.91)$$

so

$$G(x_i) = e^{-ik(z_o - z_i)}\int_{-\infty}^{\infty} \tilde{S}\left(v + \frac{\sin \alpha_R}{\lambda}\right) \tilde{F}(v)e^{2\pi iv x_i} \exp\left(2i\frac{(z_o - z_i)}{k}\pi^2 v^2\right) \, dv. \qquad (9.92)$$

Since we require that the imagery be perfect for $\tilde{S}(v) = $ const, we can see that a virtual image exists in the plane $z_i = z_o$, because in this plane $G(x_i)$ is just the Fourier transform of the Fourier transform of the object distribution, which is the object distribution itself. Hence, for $z_i = z_o$,

$$G(x_i) = \int_{-\infty}^{\infty} \tilde{S}\left(v + \frac{\sin \alpha_R}{\lambda}\right) \tilde{F}(v)e^{2\pi iv x_i} \, dv. \qquad (9.93)$$

To see the effect of limited MTF, again consider the idealized case

$$\tilde{S}(v) = 1, \qquad -v_c \leq v \leq v_c$$
$$= 0 \qquad \text{otherwise} \qquad (9.94)$$

with a single frequency object

$$F(x_o) = 1 + M \cos(2\pi v_o x_o) \qquad (9.95)$$

with Fourier transform

$$\tilde{F}(\nu) = \delta(\nu) + \frac{M}{2}\delta(\nu - \nu_o) + \frac{M}{2}\delta(\nu + \nu_o). \tag{9.96}$$

According to Eq. 9.93, then, the image is

$$G(x_i) = \int_{-\infty}^{\infty} \tilde{S}\left(\nu + \frac{\sin \alpha_R}{\lambda}\right)\tilde{F}(\nu)e^{2\pi i\nu x i}\,d\nu$$

$$= \int_{-\nu_c - \sin \alpha_R/\lambda}^{\nu_c - \sin \alpha_R/\lambda} \left[\delta(\nu) + \frac{M}{2}\delta(\nu - \nu_o) + \frac{M}{2}\delta(\nu + \nu_o)\right]e^{2\pi i\nu x i}\,d\nu$$

$$= \begin{cases} \dfrac{M}{2}e^{-2\pi i\nu_o x i}, & \dfrac{\sin \alpha_R}{\lambda} - \nu_o \le \nu_c \le \dfrac{\sin \alpha_R}{\lambda} \\[2mm] 1 + M\cos\left(2\pi\nu_o x_i\right), & \nu_c \ge \nu_o + \dfrac{\sin \alpha_R}{\lambda} \\[2mm] 0, & \nu_c \le \dfrac{\sin \alpha_R}{\lambda} - \nu_o. \end{cases} \tag{9.97}$$

These three situations are easily interpreted. In the first case, there is a uniform irradiance in the image plane. This occurs when the recording medium has been capable of recording the low-frequency sideband—carrier minus signal—but has been incapable of recording the carrier, $\sin(\alpha_R/\lambda)$. In the second case, the complete image is formed, since the medium has resolved the carrier plus both sidebands. In the third case, there is no light in the image plane since the recording medium has not even recorded the lowest frequency present.

This result points up one of the basic differences between a Fresnel hologram and the Fourier transform hologram described above. In the latter, the resolution was not affected by the limited MTF of the recording medium; only the field of view was affected. In the present case, the resolution is affected. Had the object of Eq. 9.95 contained a continuum of spatial frequencies, only those up to $\nu_{o_{max}}$ would have appeared in the image, that is, would have been resolved, where $\nu_{o_{max}} = \nu_c - \sin \alpha_R/\lambda$.

The line amplitude spread function of the holographic system will just be the image of an infinitely narrow line object. In this case, $F(x_o) = \delta(x_o)$ and $\tilde{F}(\nu) = 1$ so that Eq. 9.93 gives (again omitting e^{-2ikz_o})

$$G_l(x_i) = \int_{-\infty}^{\infty} \tilde{S}\left(\nu + \frac{\sin \alpha_R}{\lambda}\right)e^{2\pi i\nu x i}\,d\nu \tag{9.98}$$

and by Fourier inversion of Eq. 9.84, the line amplitude spread function

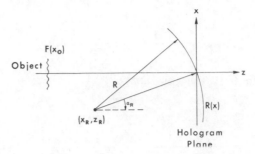

Fig. 9.12   Recording a Fresnel hologram with a spherical reference wave of radius R.

becomes

$$G_l(x_i) = \exp\left(-2\pi i\, \frac{x_i \sin \alpha_R}{\lambda}\right) S(x_i) \tag{9.99}$$

that is, just the line amplitude spread function of the recording medium, with an additional phase factor.

A still more general case that should be considered under the general topic of Fresnel holograms is that of a spherical reference wave, as indicated in Fig. 9.12. We assume that this wave has a radius R and originates from a point source located at (x_R, z_R). Its off-axis position is defined by an angle $\alpha_R = \sin^{-1}(x_R/R)$ so that, in one dimension,

$$R(x) = e^{-ik(x \sin \alpha_R + x^2/2R)}, \tag{9.100}$$

where we are assuming a uniform amplitude of unity across the hologram plane and have neglected terms of order $1/R^2$ in the exponential. For the primary image term, we return to Eq. 9.60:

$$I_1(x) = \int_{-\infty}^{\infty} S(u)[R^*(x-u)O(x-u)]\, du.$$

Substituting Eq. 9.100 for $R(x)$ and writing

$$O(x) = \left(\frac{i}{\lambda z_o}\right)^{1/2} \int_{-\infty}^{\infty} F(x_o) e^{-ikz_o} \exp\left(-ik\, \frac{(x-x_o)^2}{2z_o}\right) dx_o \tag{9.101}$$

we obtain

$$I_1(x) = \left(\frac{i}{\lambda z_o}\right)^{1/2} e^{-ikz_o} \int_{-\infty}^{\infty} S(u) e^{ik(x-u)\sin \alpha_R} e^{ik(x-u)^2/2R}$$

$$\times \int_{-\infty}^{\infty} F(x_o) \exp\left(-ik\, \frac{(x-u-x_o)^2}{2z_o}\right) dx_o\, du. \tag{9.102}$$

By interchanging the order of integration and regrouping some terms,

Eq. 9.102 may be written as

$$I_1(x) = \left(\frac{i}{\lambda z_o}\right)^{1/2} e^{-ikz_o} \int_{-\infty}^{\infty} F(x_o) \exp\left(-i\frac{kx_o^2}{2z_o}\right) \int_{-\infty}^{\infty} S(u)$$

$$\times \exp\left(i\frac{k}{2}(x-u)^2\left(\frac{1}{R}-\frac{1}{z_o}\right)\right)$$

$$\times \exp\left(ik(x-u)\left(\frac{x_o}{z_o}+\sin\alpha_R\right)\right) du\, dx_o. \quad (9.103)$$

The u-integration is now in the form of a convolution of $S(u)$ and $\beta(u)$, where

$$\beta(u) = \exp\left(ik\frac{u^2}{2R'}\right)\exp\left(iku\left(\frac{x_o}{z_o}+\sin\alpha_R\right)\right), \quad (9.104)$$

with

$$\frac{1}{R'} = \frac{1}{R} - \frac{1}{z_o}. \quad (9.105)$$

This integration will be just the Fourier transform of the product of the Fourier transforms of $S(u)$ and $\beta(u) - \tilde{S}(v)$ and $\tilde{\beta}(v)$, respectively. Now $\tilde{S}(v)$ is just the MTF of the recording medium and

$$\tilde{\beta}(v) = \int_{-\infty}^{\infty} \beta(u)e^{-2\pi ivu}\, du = (i\lambda R')^{1/2} \exp\left(-i\left(\frac{kx_o}{z_o}+k\sin\alpha_R-2\pi v\right)^2\frac{R'}{2k}\right).$$

$$(9.106)$$

Thus

$$I_1(x) = \left(\frac{i}{\lambda z_o}\right)^{1/2} e^{-ikz_o} \int_{-\infty}^{\infty} F(x_o)$$

$$\times \exp\left(-i\frac{k}{2z_o}x_o^2\right)\int_{-\infty}^{\infty} \tilde{S}(v)\tilde{\beta}(v)e^{2\pi ivx}\, dv\, dx_o$$

$$= i\left(\frac{R'}{z_o}\right)^{1/2} e^{-ikz_o} \int_{-\infty}^{\infty} F(x_o)\exp\left(-i\frac{k}{2z_o}x_o^2\right)\int_{-\infty}^{\infty} \tilde{S}(v)$$

$$\times \exp\left(-i\left(\frac{kx_o}{z_o}+k\sin\alpha_R-2\pi v\right)^2\frac{R'}{2k}\right)e^{2\pi ivx}\, dv\, dx_o. \quad (9.107)$$

Now by defining

$$\gamma \equiv k\sin\alpha_R - 2\pi v \quad (9.108)$$

and

$$\frac{1}{l} \equiv -\frac{1}{z_o} - \frac{R'}{z_o^2}\ ; \quad (9.109)$$

and doing some algebra, we find

$$
I_1(x) = \left(\frac{R'}{z_o}\right)^{1/2} \frac{e^{ik(x \sin \alpha_R - z_o)}}{2\pi i}
$$

$$
\times \int_{-\infty}^{\infty} \tilde{S}\left(\frac{\sin \alpha_R}{\lambda} - \frac{\gamma}{2\pi}\right) \exp\left(-i\frac{R'}{2k}\left(1 + \frac{lR'}{z_o^2}\right)\gamma^2\right) e^{i\gamma x}
$$

$$
\times \int_{-\infty}^{\infty} F(x_o) \exp\left[i\frac{k}{2l}\left(\frac{lR'}{kz_o}\gamma - x_o\right)^2\right] d\gamma \, dx_o. \tag{9.110}
$$

For reconstruction, let us assume that the illuminating wave is identical to the reference wave:

$$
C(x) = R(x) = e^{-ik(x \sin \alpha_R + x/2R)}. \tag{9.111}
$$

The analysis loses some generality at this point, but the resulting expressions will be more easily interpreted as to physical meaning. The amplitude distribution in the x_i plane a distance z_i from the hologram is given by

$$
G(x_i) = \left(\frac{-i}{\lambda z_i}\right)^{1/2} e^{ikz_i} \int_{-\infty}^{\infty} C(x)I_1(x) \exp\left(ik\frac{(x - x_i)^2}{2z_i}\right) dx, \tag{9.112}
$$

where we are again assuming that the amplitude transmittance of the hologram is equal to $I_1(x)$ and we are neglecting terms of the order of $1/z_i^2$ in the binomial expression of the distance between a point x in the hologram plane and the point x_i in the x_i-plane. Substitution of (9.110) and (9.111) into (9.112) yields

$$
G(x_i) = \frac{-i}{2\pi}\left[\frac{-iR'}{\lambda z_i z_o}\right]^{1/2} e^{ik(z_i - z_o)} \int_{-\infty}^{\infty} \exp\left(-i\frac{k}{2R}x^2\right) \exp\left(i\frac{k}{2z_i}(x - x_i)^2\right)
$$

$$
\times \int_{-\infty}^{\infty} \tilde{S}\left(-\frac{\gamma}{2\pi} + \frac{\sin \alpha_R}{\lambda}\right) \exp\left[-\frac{R'}{2k}\left(1 + \frac{lR'}{z_o^2}\gamma^2\right)\right] e^{i\gamma x}
$$

$$
\times \int_{-\infty}^{\infty} F(x_o) \exp\left[i\frac{k}{2l}\left(\frac{lR'}{kz_o}\gamma - x_v\right)^2\right] d\gamma \, dx_v \, dx. \tag{9.113}
$$

Now by writing $z_i = z_o$, so that we obtain the primary image, and using

$$
\int_{-\infty}^{\infty} \exp\left(-i\frac{k}{2R'}x^2\right) \exp\left[-ix\left(\frac{kx_i}{z_o} - \gamma\right)\right] dx
$$

$$
= (-i\lambda R')^{1/2} \exp\left[i\frac{R'}{2k}\left(k\frac{x_i}{z_o} - \gamma\right)^2\right], \tag{9.114}
$$

we find, after some simplification,

$$G_i(x_i) = \frac{R'}{z_0} \exp\left(-i\frac{k}{2l}x_i^2\right) \exp\left(-i\frac{kR'}{z_0}x_i \sin\alpha_R\right) \int\!\!\!\int_{-\infty}^{\infty} \tilde{S}(\nu)F(x_0)$$

$$\times \exp\left(i\frac{k}{2l}x_0^2\right) \exp\left(-i\frac{kR'}{z_0}x_0 \sin\alpha_R\right)$$

$$\times \exp\left(i\frac{R'}{z_0}2\pi\nu(x_i + x_0)\right) d\nu\, dx_0. \tag{9.115}$$

This expression tells us that, in general, when the reference wave is spherical and the object is close to the hologram plane, both the resolution and the field of view are restricted by the MTF of the recording medium. This case represents a situation midway between the situations indicated by Eqs. 9.70 and 9.93. In Eq. 9.70, for the Fourier transform hologram, the image resolution is independent of the object frequency, but the extent over which the image exists is limited by the MTF. In Eq. 9.93 the extent of the image is independent of the MTF, but the resolved object frequencies extend only up to $\nu_{0\max} = \nu_c - \sin\alpha_R/\lambda$, that is, the recording medium cutoff frequency less the carrier frequency.

To see more clearly the effect of limited MTF in the present situation, let us again consider an idealized transfer function for the medium:

$$\tilde{S}(\nu) = 1, \qquad -\nu_c \leq \nu \leq \nu_c$$
$$= 0 \qquad \text{otherwise.} \tag{9.116}$$

Then

$$G(x_i) = \frac{R'}{z_0} \exp\left(-i\frac{k}{2l}x_i^2\right) \exp\left(-i\frac{kR'}{z_0}x_i \sin\alpha_R\right) \int_{-\infty}^{\infty}\int_{-\nu_c}^{\nu_c} F(x_0)$$

$$\times \exp\left(i\frac{k}{2l}x_0^2\right) \exp\left(-i\frac{kR'}{z_0}x_0 \sin\alpha_R\right)$$

$$\times \exp\left(i\frac{R'}{z_0}2\pi\nu(x_i + x_0)\right) d\nu\, dx_0$$

$$= \frac{2\nu_c R'}{z_0} \exp\left(-i\frac{k}{2l}x_i^2\right) \exp\left(-i\frac{kR'}{z_0}x_i \sin\alpha_R\right)$$

$$\times \int_{-\infty}^{\infty} \text{sinc}\left[\frac{2\pi R'}{z_0}(x_i + x_0)\nu_c\right]$$

$$\times F(x_0) \exp\left(i\frac{k}{2l}x_0^2\right) \exp\left(-i\frac{kR'}{z_0}x_0 \sin\alpha_R\right) dx_0. \tag{9.117}$$

The line spread function for the imaging process may be found by writing

$F(x_o) = \delta(x_o)$. Doing this, we obtain

$$|G_i(x_i)|^2 = 4\left(\frac{R'}{z_o}\right)^2 v_c^2 \operatorname{sinc}^2\left[\frac{2\pi R'}{z_o} x_i v_c\right]$$

$$= 4\left(\frac{R}{z_o - R}\right)^2 v_c \operatorname{sinc}^2\left[\frac{2\pi R}{z_o - R} x_i v_c\right], \tag{9.118}$$

which has an approximate width

$$\Delta x_i = \left[\frac{z_o - R}{R}\right]\frac{1}{v_c}. \tag{9.119}$$

An interesting special case is that for $R = z_o$, that is, reference source in the object plane. Returning to Eq. 9.103, we see that the substitution $R = z_o$ eliminates the exponential that is quadratic in $(x - u)$. Thus the primary image term becomes

$$I_1(x) = \left(\frac{i}{\lambda z_o}\right)^{1/2} e^{-ikz_o} \int_{-\infty}^{\infty} F(x_o) \exp\left(-i\frac{k}{2z_o} x_o^2\right)$$

$$\times \int_{-\infty}^{\infty} S(u) \exp\left[ik(x - u)\left(\frac{x_o}{z_o} + \sin\alpha_R\right)\right] du\, dx_o. \tag{9.120}$$

Again, we note that this is a convolution integral between $S(u)$ and $\beta(u)$, where

$$\beta(u) = \exp\left[iku\left(\frac{x_o}{z_o} + \sin\alpha_R\right)\right]. \tag{9.121}$$

Without the quadratic exponential, we find

$$\tilde{\beta}(v) = \delta\left[\frac{x_o}{\lambda z_o} + \frac{\sin\alpha_R}{\lambda} - v\right] \tag{9.122}$$

so that

$$I_1(x) = \left(\frac{i}{\lambda z_o}\right)^{1/2} e^{-ikz_o} e^{ikx\sin\alpha_R}$$

$$\times \int_{-\infty}^{\infty} F(x_o)\tilde{S}\left[\frac{\sin\alpha_R}{\lambda} < \frac{x_o}{\lambda z_o}\right] \exp\left(ik\frac{x_o^2}{2z_o}\right) \exp\left(ik\frac{xx_o}{z_o}\right) dx_o. \tag{9.123}$$

Illuminating with a wave $C(x)$ given by Eq. 9.111, the primary image is given by

$$G(x_i) = \left(\frac{-i}{\lambda z_i}\right)^{1/2} \int_{-\infty}^{\infty} C(x)I_1(x)e^{ikz_i} \exp\left(i\frac{k}{2z_i}(x - x_i)^2\right) dx, \tag{9.124}$$

where we are again assuming that the amplitude transmittance of the processed hologram is simply equal to the incident irradiance. Substituting

Eqs. 9.111 and 9.123 into 9.124, and letting $z_i = z_o$ as before, we find that

$$G(x_i) = \exp\left(i\frac{k}{z_o}x_o^2\right)F(x_i)\widetilde{S}\left[\frac{\sin \alpha_R}{\lambda} + \frac{x_i}{\lambda z_o}\right]. \qquad (9.125)$$

The similarity between Eqs. 9.125 and 9.70 indicates why this type of holo-gram is referred to as a lensless Fourier transform hologram. The MTF of the recording medium, $\widetilde{S}(\nu)$, affects only the field of view, and not the resolution in the image. The same comments that applied to Eq. 9.70 also apply to Eq. 9.125.

9.4 THIRD-ORDER ABERRATIONS [4]

9.4.1 Introduction

This section treats the third-order aberrations of reconstructed wavefronts. Aberrations are introduced whenever any of the system parameters are changed between recording and reconstructing, such as wavelength, radii of curvature, or even the hologram itself. There will be no discussion of the relationship between the amount of aberration present and the image resolution; this is an enormously complicated subject in itself. The five Seidel aberrations, spherical, coma, astigmatism, field curvature, and distortion, all present in a general holographic system, are derived as a phase difference between the reconstructed wavefront of a point object and a reference sphere.

9.4.2 Analysis

Consider a point object situated at $P_o(x_o, y_o, z_o)$ in our usual coordinate system (Fig. 9.13). Let the wavelength be λ_o; then the phase of the spherical

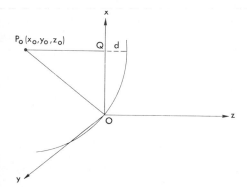

Fig. 9.13 Coordinate system for deter-mining the phase φ_o of the object wave. The object is a single point located at (x_o, y_o, z_o).

wave from P_o in the hologram, relative to the phase at the origin is

$$\varphi_o(x, y) = \frac{2\pi}{\lambda_o} d = \frac{2\pi}{\lambda_o} (\overline{P_oQ} - \overline{P_oO})$$

$$= \frac{2\pi}{\lambda_o} \{[(x - x_o)^2 + (y - y_o)^2 + z_o^2]^{1/2} - [x_o^2 + y_o^2 + z_o^2]^{1/2}\}$$

$$= \frac{2\pi}{\lambda_o} z_o \left\{ \left[1 + \frac{(x - x_o)^2 + (y - y_o)^2}{z_o^2} \right]^{1/2} - \left[1 + \frac{x_o^2 + y_o^2}{z_o^2} \right]^{1/2} \right\}.$$

$$(9.126)$$

Assuming that $z_o^2 > x_o^2 + y_o^2$, we may expand the square roots to obtain

$$\varphi_o(x, y) = \frac{2\pi}{\lambda_o} \left[\frac{1}{2z_o} (x^2 + y^2 - 2xx_o - 2yy_o) - \frac{1}{8z_o^3} (x^4 + y^4 + 2x^2y^2 \right.$$
$$- 4x^3x_o - 4y^3y_o - 4x^2yy_o - 4xy^2x_o + 6x^2x_o^2 + 6y^2y_o^2$$
$$+ 2x^2y_o^2 + 2y^2x_o^2 + 8xyx_oy_o - 4xx_o^3 - 4yy_o^3$$
$$\left. - 4xx_oy_o^2 - 4xx_o^2y_o) + \text{higher order terms} \right]. \qquad (9.127)$$

If a point source at (x_R, y_R, z_R) supplies the reference wave, also at λ_o, then a similar expression holds for the phase of the reference wave φ_R, with (x_o, y_o, z_o) replaced by (x_R, y_R, z_R). Similarly if a point source at (x_c, y_c, z_c) supplies the illuminating wave, its phase is given by the same expression with the substitution of (x_c, y_c, z_c) for (x_o, y_o, z_o), and λ_c, the illuminating wavelength, for λ_o. The phases of the primary and conjugate image waves are given by, respectively,

$$\Phi_p = \varphi_c + \varphi_o - \varphi_R \quad \text{and} \quad \Phi_c = \varphi_c + \varphi_R - \varphi_o. \qquad (9.128)$$

9.4.2.1 *Magnification*

The logical procedure is to discuss magnification here since this is calculated from the first-order terms of Eq. 9.127.

To first order in $1/z$, then, the phase of the primary image wave is given by

$$\Phi_p^{(1)} = \frac{2\pi}{\lambda_c} \frac{1}{2z_c} (x^2 + y^2 - 2xx_c - 2yy_c)$$

$$+ \frac{2\pi}{\lambda_o} \frac{1}{2z_o} (x'^2 + y'^2 - 2x'x_o - 2y'y_o)$$

$$- \frac{2\pi}{\lambda_o} \frac{1}{2z_R} (x'^2 + y'^2 - 2x'x_R - 2y'y_R). \qquad (9.129)$$

The primed coordinates refer to the coordinates of the hologram as it was recorded. A subsequent scaling up or down of the hologram results in the transformation $x = mx'$ and $y = my'$. We now define the illuminating to recording wavelength ratio $\lambda_c/\lambda_o = \mu$ and write

$$\Phi_p^{(1)} = \frac{2\pi}{\lambda_c} \frac{1}{2}\left\{(x^2 + y^2)\left[\frac{1}{z_c} + \frac{\mu}{m^2 z_o} - \frac{\mu}{m^2 z_R}\right]\right.$$

$$\left. - 2x\left[\frac{x_c}{z_c} + \frac{\mu x_o}{m z_o} - \frac{\mu x_R}{m z_R}\right] - 2y\left[\frac{y_c}{z_c} + \frac{\mu y_o}{m z_o} - \frac{\mu y_R}{m z_R}\right]\right\}. \quad (9.130)$$

We now consider Eq. 9.130 to be the first-order term of the Gaussian reference sphere defined by

$$\Phi_p^{(1)} = \frac{2\pi}{\lambda_c} \frac{1}{2}\left[\frac{x^2 + y^2 - 2xa_p - 2yb_p}{Z_p}\right] \quad (9.131)$$

with Z_p its radius and a_p and b_p the coordinates of its center, which determine the Gaussian image point. Making this necessary identification of terms, we find

$$Z_p = \frac{m^2 z_c z_o z_R}{m^2 z_o z_R + \mu z_c z_R - \mu z_c z_o} \qquad a_p = \frac{m^2 x_o z_o z_R + \mu m x_o z_c z_R - \mu m x_R z_c z_o}{m^2 z_o z_R + \mu z_c z_R - \mu z_c z_o}$$

$$(9.132)$$

with b_p given by an expression similar to that for a_p, but with the x's replaced by y's. The coordinates for the conjugate image are given by

$$Z_c = \frac{m^2 z_c z_o z_R}{m^2 z_o z_R - \mu z_c z_R + \mu z_c z_o}$$

$$(9.133)$$

$$a_c = \frac{m^2 x_c z_o z_R - \mu m x_o z_c z_R + \mu m x_R z_c z_o}{m^2 z_o z_R - \mu z_c z_R + \mu z_c z_o}.$$

If Z_p is negative, it means that the center of the reference sphere is to the left of the hologram and the image is virtual. The phase Φ_p then corresponds to a wave diverging from Z_p. If it is positive, the primary image is real and to the right of the hologram. The phase $\Phi_p^{(1)}$ then represents a spherical wave converging to Z_p. The same considerations apply to the conjugate image, so that either image may very well be real or virtual.

The magnification is given by

$$M = \frac{da}{dx_o} \quad (9.134)$$

so that

$$M_p = \frac{m}{1 + m^2 z_o/\mu z_c - z_o/z_R} \quad (9.135)$$

and

$$M_c = \frac{m}{1 - m^2 z_0/\mu z_c - z_0/z_R}.$$ (9.136)

If $Z_c \to \infty$, that is, if plane wave illumination is used, then the μ-dependence of the magnification drops out. Thus in order to achieve magnification by use of different wavelengths, one must use a spherical illuminating wave. Also, if $Z_c \to \infty$, then $M_p = M_c$.

The angular magnification for a virtual image is given by

$$M_{\text{ang}} = \frac{d(a/\mathbf{Z})}{d(x_0/z_0)} = \pm \frac{\mu}{m}$$ (9.137)

where the upper sign refers to the virtual image formed by $\Phi_p^{(1)}$ and the lower sign to the virtual image formed by $\Phi_c^{(1)}$. This latter image, then, will always be inverted, and the former always upright. If the wavelength and scale remain unchanged between recording and illuminating, the angular magnification will be unity, regardless of the other parameters, z_0, z_R, and z_c.

The longitudinal magnification of the holographic imaging process is given by

$$M_{\text{long}} = \frac{d\mathbf{Z}_c}{dz_0} = -\frac{1}{\mu} M^2$$ (9.138)

which, except for the $1/\mu$ factor, is identical to the one for conventional imaging.

9.4.2.2   Third-Order Aberrations

The third-order term of the Gaussian reference sphere is given by

$$\begin{aligned}
\Phi^{(3)} = \frac{2\pi}{\lambda_c}\Big[&-\frac{1}{8}\frac{1}{\mathbf{Z}^3}\,(x^4 + y^4 + 2x^2 y^2 - 4x^3 a - 4y^3 b - 4xy^2 a \\
&- 4x^2 y b + 6x^2 a^2 + 6y^2 b^2 + 2x^2 b^2 + 2y^2 a^2 \\
&+ 8xyab - 4xa^3 - 4yb^3 + 4xab^2 - 4ya^2 b)\Big].
\end{aligned}$$ (9.139)

The third-order term of the reconstructed wavefront is given by the sum of the third-order terms of Φ_p and Φ_c as determined from Eq. 9.156. Changing to polar coordinates ρ and θ defined by

$$\rho^2 = x^2 + y^2, \qquad x = \rho \cos \theta, \qquad y = \rho \sin \theta.$$ (9.140)

the third-order aberrations, being the phase differences between the reference sphere and the actual wavefront, separate into the five usual types:

$$\Delta\Phi = \frac{2\pi}{\lambda_c}\left\{-\tfrac{1}{8}\rho^4 S \right. \qquad\qquad\qquad\qquad\qquad \text{Spherical}$$

$$+ \tfrac{1}{2}\rho^3[C_x \cos\theta + C_y \sin\theta] \qquad\qquad \text{Coma}$$

$$- \tfrac{1}{2}\rho^2[A_x \cos^2\theta + A_y \sin^2\theta + 2A_x A_y \cos\theta\sin\theta] \quad \text{Astigmatism}$$

$$- \tfrac{1}{4}\rho^2 F \qquad\qquad\qquad\qquad\qquad\qquad \text{Field curvature}$$

$$\left. + \tfrac{1}{2}\rho[D_x \cos\theta + D_y \sin\theta]\right\}. \qquad\qquad \text{Distortion}$$

$$(9.141)$$

Table 9.1 lists the various aberration coefficients in terms of the system parameters.

Table 9.2 gives the same coefficients for the case $z_R = z_c = \infty$, that is, plane reference and illuminating waves.

Table 9.3 again gives the coefficients, but now for $z_R = z_o$, that is, reference and object points in the same plane, the so-called "lensless" Fourier transform case.

The coefficients listed in these tables are for the conjugate wavefront Φ_c; the aberrations of Φ_p are obtained simply by changing the signs of z_o and z_R. In all three cases we have listed only the x-coefficient for the off-axis aberrations; the y-coefficients are obtained simply by replacing x with y. Finally, in all expressions for D_x, we have assumed $y_c = y_R = 0$ with no loss of generality.

From Table 9.2 we see that spherical aberration is zero only if $\mu = m$, which means that the hologram is scaled according to the wavelength ratio μ. Spherical aberration disappears completely $(S = 0)$ for $z_R = z_o$ as can be seen from Table 9.1.

When $z_c = z_R = \infty$, coma may be minimized by making $\tan\alpha_c = -(\mu/m)\tan\alpha_R$. Coma will be zero then only for $\mu = m$, just as for spherical. From Table 9.3 we see that for $z_o = z_R$, coma can be eliminated by making $z_c = \pm m z_o$.

From Table 9.2, we see that astigmatism will be minimized for $\tan\alpha_c = -(\mu/m)\tan\alpha_R$, just as for coma, and A_x will disappear completely if also $\mu = m$.

For $z_R = z_o$ (Table 9.3), we see that for $x_c/z_c = -(\mu/m)(x_R/z_R)$ the expression for A_x becomes

$$A_x = \frac{1}{z_o^3}\frac{\mu}{m^2}\left[\left(\frac{1}{z_o} + \frac{\mu}{z_c}\right)(x_R^2 - x_o^2)\right]. \qquad (9.142)$$

This can only be made zero for $z_c = \mu z_o$, but this can only be done simultaneously with the condition for zero coma $(z_c = \pm m z_o)$ only if $\mu = m$.

Table 9.1

Spherical

$$S = \frac{\mu}{m^4}\left[\left(\frac{\mu^2}{m^2}-1\right)\left(\frac{1}{z_o^3}-\frac{1}{z_R^3}\right) - \frac{3\mu}{z_c}\left(\frac{1}{z_o^2}+\frac{1}{z_R^2}\right) + 3\left(\frac{m^2}{z_c^2}-\frac{\mu}{m^2 z_o z_R}\right)\left(\frac{1}{z_o}-\frac{1}{z_R}\right) + 6\frac{\mu}{z_o z_R^2 z_c}\right]$$

Coma

$$C_x = \frac{\mu}{m}\frac{1}{z_c^2}\left(\frac{x_o}{z_o}-\frac{x_R}{z_R}\right) - \frac{\mu}{m^3}\frac{1}{z_o^2}\left[\frac{x_o}{z_o}\left(1-\frac{\mu^2}{m^2}\right) + \frac{\mu}{m}\frac{x_c}{z_c} + \frac{\mu^2}{m^2}\frac{x_R}{z_R}\right]$$

$$+ \frac{\mu}{m^3}\frac{1}{z_R^2}\left[\frac{x_R}{z_R}\left(1-\frac{\mu^2}{m^2}\right)\right] - \frac{\mu}{m}\frac{x_c}{z_c} + \frac{\mu^2}{m^2}\frac{x_o}{z_o} + 2\frac{\mu}{m^2}\left(\frac{x_c}{z_c}-\frac{\mu}{m}\frac{x_o}{z_o}+\frac{\mu}{m}\frac{x_R}{z_R}\right)\left(\frac{1}{z_o z_c}-\frac{1}{z_c z_R}+\frac{\mu}{m^2}\frac{1}{z_o z_R}\right)$$

Astigmatism

$$A_x = \frac{\mu}{m^2}\frac{x_c^2}{z_c^2}\left(\frac{1}{z_o}-\frac{1}{z_R}\right) - \frac{\mu}{m^2}\frac{x_o^2}{z_o^2}\left[\frac{1}{z_o}\left(1-\frac{\mu^2}{m^2}\right)+\frac{1}{z_c}+\frac{\mu^2}{m^2}\frac{1}{z_R}\right] + \frac{\mu}{m^2}\frac{x_R^2}{z_R^2}\left[\frac{1}{z_R}\left(1-\frac{\mu^2}{m^2}\right)-\frac{\mu}{z_c}+\frac{\mu^2}{m^2}\frac{1}{z_o}\right]$$

$$+ 2\frac{\mu}{m}\left(\frac{1}{z_c}-\frac{\mu}{m^2}\frac{1}{z_o}+\frac{\mu}{m^2}\frac{1}{z_R}\right)\left(\frac{x_o x_c}{z_o z_c}-\frac{x_c x_R}{z_c z_R}+\frac{\mu}{m}\frac{x_o x_R}{z_o z_R}\right)$$

Field Curvature

$$F = A_x + A_y$$

Distortion

$$D_x = \frac{\mu}{m}\left[\left(\frac{\mu^2}{m^2}-1\right)\left(\frac{x_o^3}{z_o^3}-\frac{x_R^3}{z_R^3}\right)+\frac{x_o y_o^2}{z_o^3}\right] + \frac{3x_o}{z_o}\left(\frac{x_c}{z_c}+\frac{\mu}{m}\frac{x_R}{z_R}\right)^2 - \frac{\mu}{m}\frac{(3x_o^2+y_o^2)}{z_o^2}\left(\frac{x_c}{z_c}+\frac{\mu}{m}\frac{x_R}{z_R}\right) - \frac{3x_c x_R}{z_c z_R}\left(\frac{x_c}{z_c}+\frac{\mu}{m}\frac{x_R}{z_R}\right)$$

Table 9.2

Spherical

$$S = \frac{\mu}{m^4} \frac{1}{z_o^3} \left(\frac{\mu^2}{m^2} - 1 \right)$$

Coma

$$C_x = \frac{\mu}{m^3} \frac{1}{z_o^2} \left[\frac{x_o}{z_o} \left(\frac{\mu^2}{m^2} - 1 \right) - \frac{\mu}{m} \tan \alpha_c - \frac{\mu^2}{m^2} \tan \alpha_R \right]$$

Astigmatism

$$A_x = \frac{\mu}{m^2} \frac{1}{z_o} \left[\frac{x_o^2}{z_o^2} \left(\frac{\mu^2}{m^2} - 1 \right) + \left(\tan \alpha_c + \frac{\mu}{m} \tan \alpha_R \right)^2 - 2 \frac{\mu}{m} \frac{x_o}{z_o} \left(\tan \alpha_c + \frac{\mu}{m} \tan \alpha_R \right) \right]$$

Field Curvature

$$F = A_x + A_y$$

Distortion

$$D_x = \frac{\mu}{m} \left\{ \left(\frac{x_o^3}{z_o^3} - \tan^3 \alpha_R + \frac{x_o y_o^2}{z_o^3} \right) \left(\frac{\mu^2}{m^2} - 1 \right) + \left(\tan \alpha_c + \frac{\mu}{m} \tan \alpha_R \right) \left[3 \frac{x_o}{z_o} \tan \alpha_c + 3 \frac{\mu}{m} \frac{x_o}{z_o} \tan \alpha_R \right] \right.$$
$$\left. - \frac{\mu}{m} \left(\frac{3x_o^2 + y_o^2}{z_o^2} \right) - 3 \tan \alpha_c \tan \alpha_R \right\}$$

Table 9.3

Spherical

$$S = 0$$

Coma

$$C_x = \frac{\mu}{m}\left(\frac{x_o - x_R}{z_o}\right)\left(\frac{1}{z_c^2} - \frac{1}{m^2 z_o^2}\right)$$

Astigmatism

$$A_x = \frac{\mu}{m}\left(\frac{x_R^2}{z_o^2}\right)\left(\frac{1}{mz_o} - \frac{\mu}{mz_c}\right) + 2\frac{\mu}{m}\frac{1}{z_c z_o}\left(\frac{x_o x_c}{z_c} + \frac{\mu}{m}\frac{x_o x_R}{z_o} - \frac{x_c x_R}{z_c}\right) - \frac{\mu}{m}\frac{x_o^2}{z_o^2}\left(\frac{1}{mz_o} + \frac{\mu}{mz_c}\right)$$

Field Curvature

$$F = A_x + A_y$$

Distortion

$$D_x = \frac{\mu}{m}\left[\frac{x_o^3 - x_R^3}{z_o^3}\left(\frac{\mu^2}{m^2} - 1\right) + \frac{x_o y_o^2}{z_o^3}\left(\frac{\mu^2}{m^2} - 1\right) + \left(\frac{x_c}{z_c} + \frac{\mu}{m}\frac{x_R}{z_o}\right)\left(3\frac{x_o x_c}{z_o z_c} + 3\frac{\mu}{m}\frac{x_o x_R}{z_o^2} - 3\frac{\mu}{m}\frac{x_o^2}{z_o^2} - \frac{\mu}{m}\frac{y_o^2}{z_o^2} - 3\frac{x_c x_R}{z_c z_o}\right)\right]$$

All of the remarks applying to astigmatism also pertain to curvature of field, since the defining expressions are identical.

Table 9.1 indicates, finally, that distortion is also removed for $x_c/z_c = -(\mu/m)(x_R/z_R)$ and $\mu = m$.

We thus conclude that all of the primary aberrations will disappear simultaneously by using plane wave reference and illuminating beams of equal but opposite offset angles and by scaling the hologram in the ratio of the wavelengths. It should be remembered that this analysis applies *only* to thin holograms. An aberration-free real image may still be produced with thick holograms, although a convergent illuminating wave is required, as discussed in Chapter 8.

REFERENCES

[1] R. F. van Ligten, *J. Opt. Soc. Am.*, **56**, 1 (1966).

[2] R. F. van Ligten, *J. Opt. Soc. Am.*, **56**, 1009 (1966).

[3] A. Vander Lugt and R. H. Mitchel, *J. Opt. Soc. Am.*, **57**, 372 (1967).

[4] R. W. Meier, *J. Opt. Soc. Am.*, **56**, 219 (1966).

10 Light Sources for Holography

10.1 THE GAS LASER

By far the most common and important light source for holography today is the gas laser. We will discuss the theory of operation of gas lasers only so far as required for the production of good holograms. For an excellent, comprehensive review of gas lasers, the reader is referred to the article by Bloom [1], which also contains a large list of references.

Of the many types of gas lasers presently available, the He-Ne laser and the argon ion laser are those used for most of the hologram work being performed today. Therefore most of this discussion will center around the He-Ne laser emitting at 0.6328μ and the argon laser emitting principally at 0.4880 and 0.5145μ. The general features of these lasers discussed here are applicable to other gas lasers as well, since most of the characteristics of interest for holography pertain to properties of the resonator cavity rather than to the physics of the laser action. Primarily we are interested in the temporal and spatial properties of light.

10.1.1 Temporal Coherence

Basically, a gas laser consists of an atomic or molecular gas at low pressure contained in a long discharge tube. The axis of the discharge tube defines the optical axis of the system. Centered on this axis, at each end of the tube, is a high-reflectivity mirror, aligned so that light traveling along the axis reflects back and forth many times through the amplifying medium. A small percentage of this light is transmitted by one or both mirrors on each pass. This transmission represents a loss to the system, so for relatively low-gain lasers such as the He-Ne, high-quality, high-reflectivity mirrors are required. For higher gain lasers, such as the argon laser, the mirrors may be allowed to transmit much more light, resulting in a higher output power.

The gas is excited by means of an electrical discharge either at radio frequencies, DC, or pulsed. Under certain conditions a population inversion (i.e., more atoms in a higher-lying level than in a lower level) may be made to exist. Under these conditions the gas discharge becomes a light amplifier for the transition for which there is an inversion. A spontaneous emission causes stimulated emission, and the process cascades as the light travels along the tube. Because of the inversion, amplification of the spontaneous emission results. As the light reflects back and forth between the mirrors, the random nature of the spontaneous emissions is swamped by the in-phase, coherent nature of the stimulated emissions. The temporal coherence of the light output is quite high, but because of the nature of the amplifying transition, most laser beams are not single, but multiple frequency. Let us examine first how this multifrequency (multimode) output affects holography.

Because of the low pressure of the gas in a gas laser, the transition of interest is always Doppler broadened. This line, which would normally be an absorption line, becomes the gain envelope when an inversion exists. Thus the laser is a resonant Fabry-Perot cavity with gain. The condition for resonance is that the round-trip path between the mirrors be an integral number of wavelengths:

$$2d = m\lambda, \tag{10.1}$$

where d is the distance between the laser mirrors (cavity length), λ the wavelength at Doppler line center, and m is an integer, the axial mode number. Since $d \gg \lambda$, it is possible that the separation between adjacent allowed resonances (axial modes) may be less than the Doppler linewidth (gain envelope). If the gain is sufficiently high, more than one mode may oscillate at once, yielding a multifrequency output. The axial mode separation is determined by calculating what change in wavelength changes the axial mode number by unity:

$$\Delta\lambda_c = \frac{\lambda^2}{2d} \tag{10.2}$$

where $\Delta\lambda_c$ is the wavelength separation between cavity modes. Thus, if the laser is oscillating in a mode of wavelength λ, the adjacent allowable modes are at $\lambda \pm \Delta\lambda_c$. A typical laser spectrum is shown in Fig. 10.1. Note that for the situation shown, only three modes will oscillate since the allowable modes at $\lambda \pm 2\Delta\lambda_c$ fall below threshold. If the gain were to be increased, there are the next modes which would come into oscillation. Because of the relation (10.2), it is possible to obtain single-mode operation by shortening the laser cavity. For the He-Ne laser, $\lambda = 0.6328\ \mu$ and the Doppler width is of the order of $1.8 \times 10^{-6}\ \mu$ so that the axial mode separation equals the Doppler width for a cavity length of about 10 cm. The He-Ne system possesses sufficient gain so that single-mode lasers of this length can be made. However, the power output

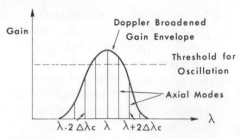

Fig. 10.1 A typical gas laser spectrum.

will be relatively small, and single-mode operation does not imply single-frequency operation, since the single mode may wander about under the gain envelope. This wandering may be caused by thermally induced changes in the cavity length, for example. In order to obtain more useful power levels, longer lasers are generally employed, resulting in a multimode output.

Since holography is basically a two-beam interference problem, consider the simple experiment shown schematically in Fig. 10.2. The laser cavity is formed by mirrors M_1 and M_2. The light transmitted by M_2 is split into two beams at the beamsplitter S, one of which travels to the point P on the recording medium, the other of which is reflected by the mirror R to the point P. The angle between the two beams is denoted by φ. These two beams interfere at the plate in the x-y plane, forming straight-line fringes lying in the x-direction. However, exposure of these fringes requires a finite exposure time and they must remain steady during this time. In general, a time-dependent wavelength from the laser, a time-dependent optical path difference $D = SRP\text{-}SP$, or a stable but multimode laser will cause demodulation of the fringes at P. Figure 10.3 shows two plane wavefronts interfering at P. For small angles, $\sin \varphi = \varphi$ and so we can write the phase of wave (1) in the x-direction as

$$\delta_1(x) = kx\frac{\varphi}{2} \tag{10.3}$$

where $k = 2\pi/\lambda$. Similarly, for wave (2) we have

$$\delta_2(x) = -kx\frac{\varphi}{2}. \tag{10.4}$$

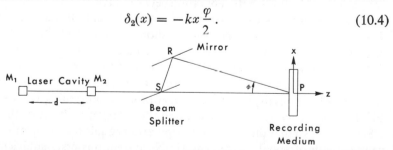

Fig. 10.2 A simple two-beam interference experiment for describing the effects of using a multimode laser for holography.

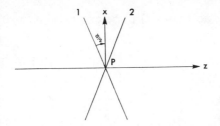

Fig. 10.3 A schematic of the wavefronts of two interfering plane waves.

In the arrangement of Fig. 10.2, one of the waves will have traveled an extra distance D and the phase difference along x will be

$$\delta(x) = k(D + x\varphi). \tag{10.5}$$

Now a bright fringe will be produced at x when $\delta(x)$ is an integral multiple of 2π. This leads to the condition

$$D + x\varphi = m\lambda \quad \text{(bright fringe)}. \tag{10.6}$$

Here m is the order of interference; its average value is D/λ. Now suppose we make a small change $\Delta\lambda$ in the wavelength. The fringes will shift an amount Δx given by

$$\Delta x = m \frac{\Delta\lambda}{\varphi}. \tag{10.7}$$

Writing $m = D/\lambda$, this becomes

$$\Delta x = \Delta\lambda \frac{D}{\lambda\varphi}. \tag{10.8}$$

But $\varphi = \lambda\nu_f$ relates the angle between the beams, the wavelength, and the spatial frequency of the fringes, ν_f, so writing $\Delta x_f = 1/\nu_f$ for the fringe spacing, we have

$$\Delta x = \frac{\Delta\lambda}{\lambda^2} D \, \Delta x_f \tag{10.9}$$

as the distance a fringe moves for a change in wavelength $\Delta\lambda$. Most commercial lasers used for holography are not completely stabilized, so small thermal variations can cause changes in the cavity length d. A change in d of $\lambda/2$ results in a wavelength shift for each mode of $\Delta\lambda = \Delta\lambda_c$. For a well-constructed laser we can expect thermal drifts to be such that $\lambda \to \lambda + \Delta\lambda_c$ in times of the order of seconds or less. In some lasers the situation is much worse, with the modes jumping about randomly and rapidly. For a change in wavelength $\Delta\lambda = \Delta\lambda_c$, the laser output spectrum is identical to that for $\Delta\lambda = 0$ and so we need not consider wavelength shifts larger than $\Delta\lambda_c$. The effect of these shifts on the interference pattern can be analyzed as follows:

suppose the path difference between the two beams D to be expressed as a fraction of the cavity length d:

$$D = Kd. \tag{10.10}$$

Then we can write

$$\Delta x = \frac{\Delta \lambda_c}{\lambda^2} Kd \, \Delta x_f, \tag{10.11}$$

which expresses the change in position of a fringe for a wavelength shift $\Delta \lambda_c$ and path difference Kd. But $\Delta \lambda_c = \lambda^2/2d$, so

$$\Delta x = \tfrac{1}{2} K \, \Delta x_f. \tag{10.12}$$

Thus we see that an expansion of the laser cavity by $\lambda/2$ causes the output spectrum to shift by $\Delta \lambda_c$ resulting in a fringe displacement Δx. For a path difference between the two beams equal to the cavity length D $(K = 1)$, the fringes shift by an amount $\Delta x_f/2$. If this happens during one exposure time, the fringe pattern on the photographic plate will be essentially completely demodulated, that is, no fringes will be recorded.

An important point to note about this result is that if we express the fringe shift as a fraction of the fringe spacing,

$$\frac{\Delta x}{\Delta x_f} = \tfrac{1}{2} K, \tag{10.13}$$

we see that it is independent of the spatial frequency of the fringe pattern; high-frequency fringes (large φ) are no more sensitive to wavelength variation than low-frequency fringes.

Next let us assume that the laser is operating at a single wavelength that does not drift in time (single mode, stabilized) and examine the effect on the interference fringes of a time-varying optical path difference. Recall from Fig. 10.2 that the two beams of light travel different paths to the photographic plate. A time-dependent path difference can be introduced by unequal temperature in the two arms which change in time, slight air drafts in the room, or by vibration of the mirror and beam splitter supports. We can analyze these effects by referring to Eq. 10.6:

$$D + x\varphi = m\lambda \quad \text{(bright fringe).}$$

A small change in D, denoted by ΔD, results in a change in fringe position Δx given by

$$\Delta x = \frac{\Delta D}{\varphi} \tag{10.14}$$

where we are neglecting the minus sign and assuming that λ and m remain constant. Writing

$$\varphi = \lambda \nu_f = \frac{\lambda}{\Delta x_f} \qquad (10.15)$$

we have

$$\Delta x = \frac{\Delta x_f \Delta D}{\lambda}. \qquad (10.16)$$

Hence we see that we have essentially complete demodulation of the fringes for $\Delta D = \lambda/2$. We also see that the fractional fringe shift, $\Delta x/\Delta x_f$, is independent of the frequency of the fringes. Here again, the sensitivity of the modulation of the fringe system to path variations is no greater for high-frequency fringes than for low-frequency fringes.

Next let us examine how the modulation (visibility) of the fringe pattern is affected by a multimode laser. We will assume that both the mode position and the path difference are stationary in time.

A gas laser of sufficient gain oscillates in more than one longitudinal (axial) mode if the separation of these modes is smaller than the width of the atomic line of the transition involved. The wavelength separation of adjacent axial modes is given by Eq. 10.2:

$$\Delta \lambda_c = \frac{\lambda^2}{\Delta \lambda}.$$

The modulation M of an interference pattern may be defined as

$$M \equiv \frac{I_{\max} - I_{\min}}{I_{\max} + I_{\min}} \qquad (10.17)$$

where $I_{\max}$ and $I_{\min}$ are, respectively, the values of the irradiance at the maxima and minima of the fringe pattern. This modulation is related to the spectral intensity distribution of the light source through a Fourier transform. A good treatment of this relationship may be found in Born and Wolf [2]. Here it is shown that the modulation of the fringes of two interfering beams of equal irradiance can be written

$$M = \frac{(S^2 + C^2)^{1/2}}{P}, \qquad (10.18)$$

where,

$$S(D) \equiv 2 \int j(x) \sin (xD) \, dx \qquad (10.19)$$

$$C(D) \equiv 2 \int j(x) \cos (xD) \, dx \qquad (10.20)$$

$$P \equiv 2 \int j(x) \, dx. \qquad (10.21)$$

Here D is the optical path difference as before, $j(x)$ is the spectral intensity distribution of the light source, and

$$x = k - k_o \tag{10.22}$$

where $k = 2\pi/\lambda$ and $k_o = 2\pi/\lambda_o$, λ_o being an arbitrarily chosen wavelength near the center of the Doppler line.

A good approximation to a multimode gas laser source is that $j(x)$ is a sum of Dirac delta functions $\delta(x)$, with each spike separated from its nearest neighbor by

$$\Delta k = \frac{2\pi}{2d}. \tag{10.23}$$

If we also assume that the spectral distribution is symmetrical about line center, then $S(D) = 0$ and the fringe modulation can be written

$$M = \frac{|C(D)|}{P}. \tag{10.24}$$

For the delta-function spectrum, we have, for a single mode,

$$j_1(x) = \delta(x) \tag{10.25}$$

and for two modes

$$j_2(x) = \delta\left(x - \frac{\Delta k}{2}\right) + \delta\left(x + \frac{\Delta k}{2}\right) \tag{10.26}$$

and for three modes

$$j_3(x) = \delta(x - \Delta k) + \delta(x) + \delta(x + \Delta k) \tag{10.27}$$

and so on. Using these expressions for the $j_n(x)$, we obtain for the fringe modulation M_n, where n denotes the number of modes oscillating,

$$M_1(D) = 1$$

$$M_2(D) = \frac{|2\cos\left(\tfrac{1}{2}\Delta kD\right)|}{2} = \left|\cos\left(\frac{\pi D}{2d}\right)\right|$$

$$M_3(D) = \frac{|2\cos(\Delta kD) + 1|}{3} = \frac{|2\cos(\pi D/d) + 1|}{3}$$

$$M_4(D) = \frac{|2\cos\left(\tfrac{3}{2}\Delta kD\right) + 2\cos\left(\tfrac{1}{2}\Delta kD\right)|}{4}$$

$$= \frac{|\cos(3\pi D/2d) + \cos(\pi D/2d)|}{2}$$

$$M_5(D) = \frac{|2\cos(2\Delta kD) + 2\cos(\Delta kD) + 1|}{5}$$

$$= \frac{|2\cos(2\pi D/d) + 2\cos(\pi D/d) + 1|}{5}$$

and so on. Evaluating these for path differences as equal to an integral multiple of the cavity length yields

$$M_n(o) = 1 \qquad \text{regardless of } n$$
$$M_n(d) = n^{-1} \quad n \text{ odd}$$
$$\qquad = 0 \qquad n \text{ even}$$
$$M_n(2d) = 1 \qquad \text{regardless of } n.$$

Thus for zero path difference the fringe modulation is 100% no matter how many modes are oscillating. For a path difference equal to one cavity length d, the fringe modulation depends on the number of modes; for example, if three modes are oscillating, the modulation is $\frac{1}{3}$, assuming all have equal strength. In any actual laser, of course, the modes may not be of equal strength and will have to be weighted accordingly. As a typical example, suppose the three modes are in the ratio $1:2:1$. Then $P = 4$ and D $(D = d) = 2 \cos \Delta k d + 2$ and so $M_3(d) = 0$ and not $\frac{1}{3}$. Thus the modulation of the fringes at $D = d$ will generally be less than n^{-1} for an odd number of modes.

In summary, we note that we have examined the problem of forming interference fringes with gas laser light for three cases: (1) a time-varying wavelength in the laser output, (2) a time-varying optical path difference between the two interfering beams, and (3) a multimode but stable laser output. All three of these perturbations can cause a serious demodulation of the fringes with a resultant loss in diffracted flux, or possibly even a complete loss of the hologram. These effects can, of course, occur singly or in any combination. In all cases, the degree of demodulation was found to be independent of the spatial frequency of the fringes. Problem (1) can be eliminated by working near zero path difference, (2) by stabilization of the supports and by blocking possible air drafts, and (3) by working near zero path difference or a path difference that is an even multiple of the cavity length.

A nonzero spectral bandwidth of the source used for illuminating the hologram mainly affects the resultant image resolution, as discussed previously. Because of this, the gas laser still represents the optimum illuminating source because of its high power in a narrow bandwidth. Even though the effective bandwidth of a multimode laser must be taken as the full Doppler width, this is still usually narrower than most spectral sources.

10.1.2 Spatial Coherence

The spatial coherence of the light used in holography is important in two respects. In recording the hologram, both reference and object beams must have well-defined wavefronts that are constant in time. The object wave may, of course, be highly complex in shape, but should nevertheless be constant.

Also, in reconstructing the object wave, the resolution in the final image depends on the size of the source of the illuminating wave, as discussed in Chapter 9.

The spatial coherence of a light field is a measure of the degree of phase correlation at two different points of space at a single time. This correlation is primarily dependent on the size of the source from which the light originated, and to find this it is necessary to know the precise spatial distribution of flux from a gas laser.

The lowest-order (spatial) mode from a gas laser has an irradiance variation of the form

$$I(\rho) = I_o e^{2\rho^2/w^2} \tag{10.28}$$

where ρ is the radial distance from the axis and w is called the spot size, that is, the value of ρ at which the irradiance has fallen to $1/e^2$ of its central value. The wavefront is either spherical or plane. Suppose we wish to focus the laser light with this distribution to a small point with a lens. The spot size at focus is given by

$$w_o = \frac{\lambda}{\pi \text{N.A.}} \tag{10.29}$$

where λ is the wavelength of the light and N.A. is the numerical aperture of the lens used to focus the light, which in this case is w/f, where f is the focal length of the lens. If w is larger than the diameter of the focusing lens, then the focused spot size will be just the Airy disk radius. The irradiance distribution at focus is thus given by

$$I(\rho) = I_o e^{-2\rho^2/w_o^2} = I_o e^{-2\pi^2 \rho^2 (\text{N.A.})^2/\lambda^2}. \tag{10.30}$$

For $\lambda = 0.6328\ \mu$ (He-Ne laser), the spot size as a function of the N.A. of the focusing objective is

$$w_o = \frac{0.2}{\text{N.A.}}\ \text{microns} = \frac{0.2f}{w}\ \text{microns} \tag{10.31}$$

assuming that the spot size w of the light entering the lens is smaller than the lens diameter. This is the spot size that will determine the resolution in the reconstruction, as described in Chapter 9.

For the case in which more than a single transverse mode is oscillating, the situation worsens. Not only is there a smaller degree of spatial coherence, but the beam cannot be focused down to as small a spot as in the single-mode case. Thus, in general, whenever a single, higher-order mode, or several higher-order transverse modes are oscillating, both the diffraction efficiency (because of a smaller fringe contrast) of the hologram *and* the resolution in the image will suffer. Fortunately virtually all commercially available lasers are designed to operate in a single, low-order transverse mode.

10.2   THE SOLID-STATE LASER

The solid-state laser most often used for holographic applications is the pulsed ruby laser, emitting at a wavelength just slightly longer than that of the He-Ne gas laser, 0.6943 μ. The fact that a ruby laser is pulsed has some very definite advantages over a continuous source for some aspects of holography. In particular, because of the short pulse duration, some subject movement can be tolerated during an exposure. A hologram made with a short pulse of light from a ruby laser can record the object at one instant of time and later the resulting three-dimensional image can be studied in detail.

The major drawback to making a hologram with a pulsed laser is the limited coherence length of most solid-state lasers. Single axial mode He-Ne lasers are available commercially, but have low power output. Even multi-mode gas lasers can be used with path differences of the order of 10–20 cm, while with most pulsed lasers one is limited to path differences of the order of 1 cm. Pulsed ruby lasers can be operated in a single axial mode, thus competing with gas lasers for coherence length, by putting additional reflec-tors in the laser cavity. Hercher [3] has described this method for obtaining single-mode operation of a Q-switched ruby laser. McClung and Wiener have improved the temporal coherence with axial mode control [4].

Another drawback to the use of the pulsed ruby laser as a source for holography is the fact that the wavefront of the output is not simply defined. The He-Ne gas laser generally operates in the single, lowest-order spatial mode, to so-called TEM_{00} mode, having the intensity distribution given in 10.28. On the other hand, it is difficult to describe the exact spatial distribution of the output of a pulsed ruby laser. Jacobson and McClung have improved the spatial distribution of a ruby laser, but at the expense of total energy in the pulse [5]. In view of the photographic sensitivity problem mentioned above, this energy loss can become quite serious. However, they were able to make holograms with such a source.

Brooks et al. [6] have successfully produced holograms with a pulsed ruby laser without mode control. This has been done by a careful choice of the optical arrangement so as to minimize the effects of the relatively low temporal and spatial coherence of ruby laser light. This is accomplished by arranging the holographic system for zero path difference between object and reference beams and also by matching as closely as possible the relative spatial positions of the two wavefronts. When these two conditions are met, the two waves will be coherent with each other and a good hologram will be recorded.

A simple arrangement that has been used successfully with the ruby laser is shown in Fig. 10.4. The zero optical path condition is easily met by equalizing the optical paths in each arm. The spatial distribution is matched in the two

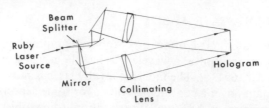

Fig. 10.4 A simple arrangement for spatially matching two interfering waves [6].

arms by expanding both beams to the same size and reflecting each the same number of times. The optical path difference between the two beams increases with increasing distance from the center of the hologram, since the order of interference is increasing. This, as with Gabor holograms, limits the effective aperture of the hologram. Brooks et al. [6] have compensated for this effect by using a stepped retardation plate, such as shown in Fig. 10.5. With this arrangement, the ray passing through the center of each step travels the same optical path as the corresponding ray in the other beam. The disturbing effect of diffraction at the edges of the step is minimized with the use of a diffuse object beam. However, the use of a diffuser causes a serious spatial mismatch of the two wavefronts, resulting in a loss in fringe contrast. Any point on the reference wave must interfere with a point on the object wave whose amplitude and phase will be determined by a sum of components from each point of the original wave. Hence, unless there is a large degree of spatial coherence, the introduction of a diffusing plate into one of the beams will reduce the fringe contrast. For a typical ruby laser, then, only a limited diffusing angle can be recorded.

Brooks et al. have also described a more sophisticated optical arrangement that allows the use of a diffuser with large diffusing angles without seriously upsetting the spatial match. Their arrangement is shown in Fig. 10.6. With this arrangement, the diffuser is imaged into the hologram plane. This creates

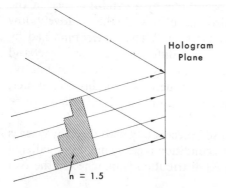

Fig. 10.5 A simple path length compensation scheme [6].

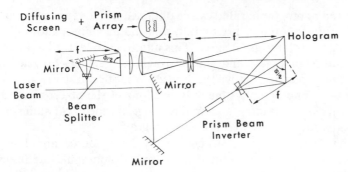

Fig. 10.6 A scheme for matching wavefronts and also allowing the use of a diffuser [6].

a one-to-one correspondence between a point on the object wavefront with the corresponding point on the reference wavefront. Another lens, identical to the one imaging the diffuser onto the hologram plane (a pair of planoconvex lenses), is placed next to the diffuser and directs the diffuse light toward the imaging lens. Also, this lens compensates for the increasing path lengths at increasing distances from the center of the diffuser. If the reference beam is incident on the hologram plane at an angle $\varphi/2$, then the path length increases from one side of the plane to the other; this is compensated for in the object beam by directing the light onto the diffuser at an angle $\varphi/2$ also. This light is then redirected toward the recording plane by means of an array of small prisms, making optimum use of the light.

This arrangement, though somewhat cumbersome, allows spatial and temporal matching of the two beams to within the limits of the lens aberrations. This is perfectly suitable for making holograms of diffusely illuminated objects with a ruby laser source. It has been tested by the authors of [6] with the object placed midway between the diffuser and the imaging lens. The recorded object was observable over about 40° with reasonably uniform diffuse illumination.

With the foregoing limitations in mind, we now proceed to discuss the prime advantage in using a ruby laser as a source for holography, namely, the short pulse duration. Assuming that enough energy to expose the recording medium is emitted in a single pulse, the total holographic exposure is made in a very short time. This means that normal vibration and motion tolerances can be very much relaxed. As a rule of thumb, object motion of about $\lambda/2$ during an exposure will obliterate the hologram because of the resultant motion of the fringes. Since the pulse duration of a typical ruby laser is of the order of 1 msec, the maximum allowable subject velocity is about 0.35 mm/sec. This is quite an appreciable subject motion when compared with the maximum allowable velocity for a 1-min exposure with a gas laser: 5×10^{-6} mm/sec. But this is still not the ultimate, since sufficient energy for a holographic exposure (on photographic film) can be obtained with a Q-switched ruby laser. Pulse durations of 30 nsec are typical, and the maximum allowable subject velocity is now as large as 10 m/sec. This number can be further increased by suitable choice of object and geometrical arrangement. For example, an object moving in silhouette will be subject to motion restrictions essentially the same as in photography: the only portions of the object affected by the motion will be the leading and trailing edges, which will suffer a loss in resolution exactly as in conventional photography.

The combination of a Q-switched ruby laser source and holographic interferometry is a very powerful analytic tool. Holographic interferometry is greater detail in Chapter 12; it can be described simply as follows. The hologram is double-exposed with two objects. Subsequent illumination reconstructs the two object waves simultaneously, and if the two objects were similar enough (in practice, the same object slightly deformed or displaced is used for the second object) these two waves will interfere and fringes will be visible. With a Q-switched ruby laser source and a diffuse, moving object, a three-dimensional interferogram of the object in motion can be made. Many fine examples of holographic interferometry with transient phenomena have been demonstrated by Heflinger, Weurker, and Brooks [6, 7]. A few examples of their work are shown in Figs. 10.7–10.9.

The expense (ruby laser systems are generally more expensive than gas lasers) and limited coherence of the ruby laser have prevented its widespread use as a holographic source. The coherence problems can be overcome with suitable arrangements, and in view of the important advantage of the ultra short exposure time, the ruby laser should become a very important source for holography.

10.3   THE LINE SOURCE

Since the advent of lasers, very few holograms have been made with a non-laser source. Of course, all of the holograms made prior to about 1962 were made with nonlaser sources, but the problems were many and the rewards few. Since the cost of a laser is now so small, it is hardly worth the effort to produce holograms with a nonlaser source. However, as was mentioned in Chapter 1, the widespread interest in holography today stems not only from the invention of the laser but also from such innovations as the offset reference beam and diffuse illumination. The first off-axis holograms were made with a filtered mercury arc as a source, as were the very first holograms in 1948. The The off-axis reference beam hologram requires that some 10^5 interference orders be recorded for a 30° angle between the beams and a 10-cm hologram. Recording this many orders requires a very narrow-band source, the maximum allowable bandwidth being only of the order of 0.06 Å. Most arc sources are

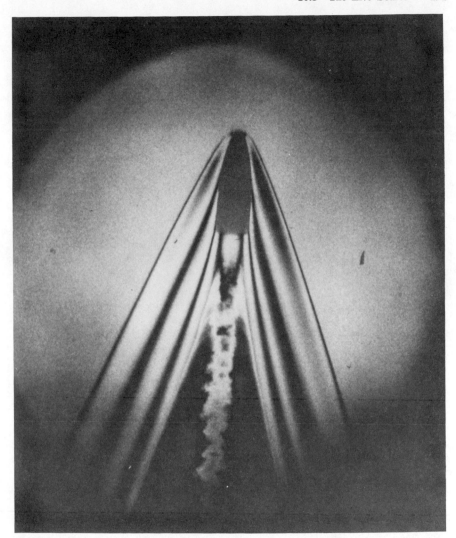

Fig. 10.7   A holographic interferogram of the shock wave of a 0.22 caliber bullet made with a pulsed ruby laser [7]. (Courtesy of TRW, Inc.)

not this monochromatic, and the few that are have a very low output power. There are, however, several methods for reducing the coherence requirements. These are generally interferometer arrangements yielding achromatic fringes [6, 8].

Achromatic fringe systems are formed when the phase difference between the two interfering beams is independent of wavelength. These are formed using a dispersing element such as a grating or prism, as described by

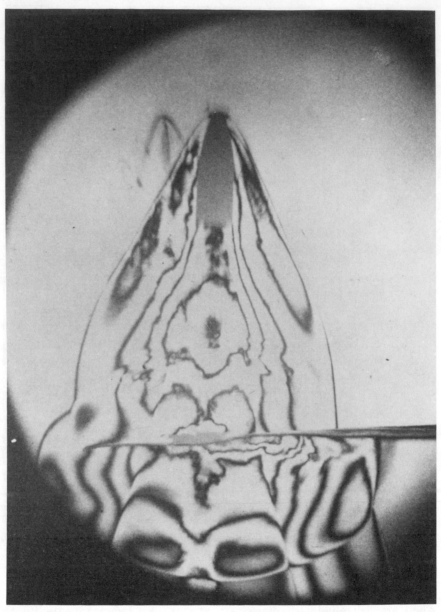

Fig. 10.8 A holographic interferogram of the shock wave of a bullet that has just passed through a sheet of 0.04-mm brass shim stock [7]. (Courtesy of TRW, Inc.)

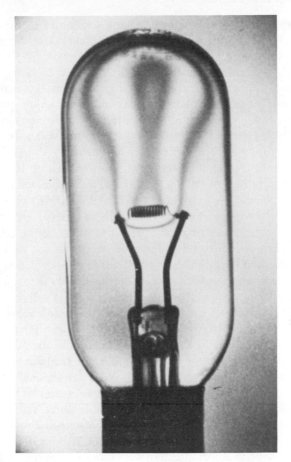

Fig. 10.9 A holographic interferogram of the change in gas distribution inside a light bulb [7]. (Courtesy of TRW, Inc.)

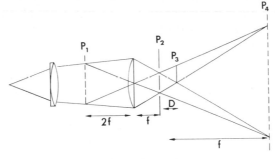

Fig. 10.10 An arrangement for recording an off-axis hologram with a line source [10].

Ditchburn [9]. Using such a fringe system, Leith and Upatnieks [10] have produced holograms with a high-pressure mercury arc, though of a lower quality than those made with a laser source.

Their method is best explained using the diffraction grating model for the off-axis type of hologram. Since the hologram represents a modulated diffraction grating, it can be made by imaging a diffraction grating onto the hologram recording medium and modulating this image so that it becomes a hologram. The coherence requirements on the source will be reduced, since an image of the grating will form even in white light.

A suitable arrangement used by them is shown in Fig. 10.10. The second lens images the grating onto plane P_4. Parallel rays diffracted from the grating are imaged at plane P_2. A spatial filter in this plane allows only two orders to pass on to form the grating image. One of these orders is used as a reference beam and the other as an object beam. That the resulting fringe system in plane P_4 is achromatic may be understood best from the viewpoint that the fringes are images of the grating rulings. The fringes can also be considered as result of interference between the two orders passed by the stop in plane P_2. If these are the two first orders, then the recorded fringes will have just twice the spatial frequency of the grating with a sinusoidal irradiance distribution.

It is easily shown that the required source coherence is lowest for the in-line holograms first described by Gabor. This is because the angle between the scattered (object) wave and the background (reference) wave is smaller (for a hologram of a given size) than for an off-axis arrangement. This small angle implies a small path difference between the two beams at the edges of the hologram, and hence less source coherence is required than for an off-axis hologram. For the achromatic fringe system described above, consider first that the object is placed in one beam in plane P_3 (Fig. 10.10). The waves scattered by the (thin) object interfere with the coherent background, and the recorded distribution in plane P_4 reflects in an in-line hologram, with the same requirements on source coherence.

If we now allow another order to be transmitted through plane P_2 to P_4, the wave scattered by the object interferes with this beam to form the grating image. This interference will take place over the same distance along P_4 that the background wave interfered with the scattered wave. However, many more fringes (orders) are present because of the larger angle between the two beams. Thus, the achromatic fringe hologram method allows off-axis holography with exactly the same source coherence requirements as for in-line holography. All of the usual benefits of off-axis holograms are preserved, such as separation of the conjugate and primary images, relative insensitivity to the nonlinearities of the recording medium, avoidance of the overlapping flare light caused by the self-interference of different object points, and the possibility of using beam balances near unity.

REFERENCES

[1] A. L. Bloom, *Appl. Opt.*, **5**, 1500 (1966).

[2] M. Born and E. Wolf, *Principles of Optics*, Pergamon Press Ltd., London, 1959, p. 319.

[3] M. Hercher, *Appl. Phys. Letters*, **7**, 39 (1965).

[4] F. J. McClung and D. Wiener, *IEEE J. Quant. Elec.*, **QE-1**, 94 (1965).

[5] A. D. Jacobson and F. J. McClung, *Appl. Opt.*, **4**, 1509 (1965).

[6] R. E. Brooks, L. O. Heflinger, and R. F. Wuerker, *IEEE J. Quant. Elec.*, **QE-2**, 275 (1966).

[7] R. E. Brooks, R. F. Wuerker, L. O. Heflinger, and C. Knox, Paper presented at the International Colloquium on Gasdynamics of Explosions, Brussels, Belgium, September 20, 1967.

[8] J. M. Burch, J. W. Gates, R. G. N. Hill, and L. H. Tanner, *Nature*, **212**, 1347 (1966).

[9] R. W. Ditchburn, *Light*, Blackie and Son, Ltd. London, 1953, pp. 145–148.

[10] E. N. Leith and J. Upatnieks, *J. Opt. Soc. Am.*, **57**, 975 (1967).

11 Recording Materials for Holography

11.0 INTRODUCTION

The problem of developing an ideal recording medium for holography has occupied much of the total research time that has gone into holography. This is because there is no universal recording material that is suitable for all of the varied applications for holography. Each application has its own unique set of requirements, and therefore there are many different materials that are suitable for recording holograms.

The very first holograms were recorded by Gabor on photographic films and plates, and, indeed, the photographic emulsion is still the principal recording material for holography. The reason for its continued widespread use, aside from its familiarity, is that the photographic emulsion comes close to having ideal properties for a wide range of requirements. However, there are some applications for which the photographic emulsion is appreciably less than ideal, and this is the reason for the intensive search for new materials.

11.1 GENERAL REQUIREMENTS

1. Resolution. The holographic recording material must be at least capable of recording the basic fringe pattern. In optical holography, where the wavelength is short, even the slightest offset angle between the reference and object beams results in the formation of a very high spatial frequency fringe pattern at the recording plane. Recall from Fig. 3.8 that the spatial frequency to be recorded is given by $(\sin \phi)/\lambda$. Thus for $\lambda = 0.5\ \mu$, an angle as small as 30° results in a frequency of about 1000 cycles/mm, which is considerably finer detail than most recording materials are capable of resolving. Thus special high-resolution materials are required in general, except in a few instances, such as recording in-line Gabor holograms.

In general, an in-line Gabor hologram requires the least amount of spatial frequency bandwidth, and an off-axis diffuse-subject hologram requires the most. For the in-line type, limited resolution of the recording medium results directly in a loss of image resolution. Generally speaking, an off-axis hologram image suffers only a loss of field for limited recording medium resolution. This becomes an exact statement for Fourier transform holograms.

It is easy to show that the spatial frequency band that is to be recorded is the same as that of the object. Consider an object transparency $F(x_o, y_o)$ located on the hologram axis a distance z_o from the hologram (x, y) plane. The field at the hologram is given by Eq. 3.2:

$$O(x, y) = \frac{-i}{2\lambda} \int\int_A F(x_o, y_o)[1 + \cos\theta] \frac{e^{ikr}}{r} \, dx_o \, dy_o \qquad (11.1)$$

where $k = 2\pi/\lambda$ and λ is the recording wavelength. Assuming θ to be small, we write $1 + \cos\theta \approx 2$, $r \approx z_o$ in the denominator and $r = [z_o^2 + (x - x_o)^2 + (y - y_o)^2]^{1/2}$ in the exponent and obtain

$$O(x, y) = \frac{-i}{\lambda z_o} \int\int_A F(x_o, y_o) \exp\{ik[z_o^2 + (x - x_o)^2 + (y - y_o)^2]^{1/2}\} \, dx_o \, dy_o.$$

$$(11.2)$$

Using the binomial expansion for the square root in the exponent, we have

$$O(x, y) = \frac{-i}{\lambda z_o} \int\int_A F(x_o, y_o) \exp\left\{ik\left[z_o + \frac{(x - x_o)^2}{2z_o} + \frac{(y - y_o)^2}{2z_o}\right]\right\} \, dx_o \, dy_o.$$

$$(11.3)$$

We can write this as a convolution:

$$O(x, y) = F * g(x, y) \qquad (11.4)$$

where

$$g(x, y) = \frac{-i}{\lambda z_o} \exp\left\{ik\left[z_o + \frac{x^2}{2z_o} + \frac{y^2}{2z_o}\right]\right\}. \qquad (11.5)$$

Since a convolution of two functions corresponds to a product of their Fourier transforms, we can write

$$\tilde{O}(\alpha, \beta) = \tilde{F}(\alpha, \beta) \cdot \tilde{g}(\alpha, \beta) \qquad (11.6)$$

where the tilde implies a Fourier transform and α and β are spatial frequency variables corresponding to x and y. Now

$$\tilde{g}(\alpha, \beta) = \frac{-i}{\lambda z_o} e^{ikz_o} e^{i(z_o/2k)(\alpha^2 + \beta^2)} \qquad (11.7)$$

so that $|\tilde{g}(\alpha, \beta)| = $ const. This means that the power spectrum of O and F are the same; both have the same spatial frequency content. Therefore, recording the signal wave requires no more resolution than does recording the image of the transparency.

This all pertains only to the recording of in-line holograms. The situation is different for an off-axis hologram, since in the latter case the diffraction pattern to be recorded is modulated onto a spatial carrier frequency. Hence the resolution requirements on the recording medium are approximately doubled. Since the interference is usually between two beams whose axes define a plane, the doubling of the resolution requirements is in only one dimension; the requirements for the other dimension remain unchanged.

To record the full field of view, the recording medium must resolve the spatial frequencies generated by the interference of the light from each object point with the reference beam. This requirement is different for the various hologram types.

A. FOURIER TRANSFORM HOLOGRAMS.   For an object of height $2A$ and a lens of focal length f, the range of spatial frequencies at the hologram is given by

$$\Delta \nu = \frac{2}{\lambda} \sin \left(\tan^{-1} \frac{A}{f} \right).$$
(11.8)

Hence the range of spatial frequencies that the recording medium must resolve depends only on the object size.

B. FRENSEL HOLOGRAMS.   The range of frequencies to be recorded, for a plane reference wave, is given by

$$\Delta \nu = \frac{\sin \{\tan^{-1} [(H - A)/z_o]\} + \sin \{\tan^{-1} [(H + A)/z_o]\}}{\lambda},$$
(11.9)

where the height of the hologram is $2H$ and z_o is the object distance. Here the resolution requirement depends on both the object and hologram size.

C. LENSLESS FOURIER TRANSFORM HOLOGRAMS.   For this type of hologram, the highest spatial frequency occurs near the center of the hologram. Usually the hologram is recorded with the object and reference source displaced equally from the axis. If the center of the object is at a height A above the axis, then

$$\Delta \nu = \frac{\sin \{2(A/z_o)[1 - (A^2/3z_o) + \cdots]\}}{\lambda}.$$
(11.10)

Each object point yields an essentially constant spatial frequency on the recording medium, and we have an approximation to the Fourier transform case.

2. Sensitivity. The hologram recording material must be capable of recording the fringe pattern within a reasonable time. In order to do this it must possess sufficient sensitivity to the particular radiation involved. Since there is such a wide range of wavelengths over which people are interested in making holograms (x-rays to microwaves), there is also a wide range of acceptable sensitivities. As a general rule, however, for the visible range of wavelengths, the required exposure should be less than about 10^4 ergs/cm². For the sources that are available today, a larger exposure requirement would lead to an intolerably long exposure time. It should be pointed out, however, that there are certain special cases, such as recording small holograms with a powerful laser source, for which a recording material having an energy requirement in excess of 10^4 ergs/cm² would be perfectly adequate. The most desirable range for recording a typical 4 × 5-inch hologram with visible light would be about 100 ergs/cm².

3. Spectral Range. The hologram recording material must be capable of recording the spectral range of interest. Most materials have some intrinsic sensitivity in the blue, so that this spectral region is generally not a problem. As the range is extended to the green and red this intrinsic sensitivity falls off and some form of spectral sensitizer must be used. This spectral sensitization has been accomplished most successfully with silver halide photographic emulsions, and so when recording holograms in the red and far red these materials are used almost exclusively.

4. Other. Other general requirements that might be considered are such things as ease of handling and ease of processing. The first of these, ease of handling, is determined mainly by the physical characteristics of the material, such as whether or not it is on a flexible support, for example, a film, or whether it is in the form of a crystal. Ease of processing is an important consideration because complex processing procedures are lengthy and time consuming, and so should be avoided.

11.2 HOLOGRAM TYPES

The hologram recording material that is to be used depends on the type of hologram to be recorded. Holograms are classified according to the type of modulation they impose on the illuminating wave (amplitude or phase), the thickness of the recording material relative to the fringe spacing (thick or thin), and the mode of image formation (transmission or reflection).

The amplitude transmittance of a material is a complex function that describes the change in the amplitude and phase of an incident light wave upon transmission through the material. If the incident wave is described by the complex function $A_i(x) = b_i(x)e^{i\phi_i(x)}$ and the transmitted wave at the point x by the function $A_t(x) = b_t(x)e^{i\phi_t(x)}$, then the amplitude transmittance of the material is defined by the ratio of these two quantities:

$$T_a(x) = \frac{A_t(x)}{A_i(x)} = \left[\frac{b_t}{b_i}\right]e^{i(\varphi_t - \varphi_i)}. \tag{11.11}$$

For the common case of an incident plane wave, A_i is a constant and the complex amplitude transmittance is given simply by

$$T_a(x) = b_t(x)e^{i\varphi_t(x)}. \tag{11.12}$$

If $b_t(x)$ is not a constant but $\phi_t(x)$ is, it is the amplitude of the incident wave that is modulated and the hologram is said to be an amplitude hologram. The magnitude of $b_t(x)$ depends on the absorption constant a_o and the thickness of the material d according to the equation

$$b_t(x) = b_i(x)e^{-a_o(x)d} \tag{11.13}$$

(see Eq. 7.21).

For an amplitude hologram the transmitted amplitude of the illuminating wave is modulated in accordance with the exposure. This is accomplished causing a spatial variation in the absorption constant a_o. The two most common materials that can accomplish this as silver halide photographic emulsions and photochromic glasses.

From Eq. 11.13 we see that the relationship between amplitude transmittance and absorption constant is

$$T_a = e^{-a_o d} \tag{11.14}$$

and in terms of the optical density

$$D = -\log_{10}(T_a^2) = 0.869a_o d \tag{11.15}$$

If the hologram is thin (see Eq. 7.1) then the optimum average amplitude transmittance, theoretically, is 0.5. On the other hand, if the hologram is thick then the optimum average amplitude transmittance is about 0.4, corresponding to an optical density of around 0.8.

The most common recording material for amplitude holograms is the photographic emulsion, simply exposed and processed in the normal manner so that the silver density absorbs and modulates the illuminating light. The thickness of these materials range from a few microns to almost 20 μ for those that are commercially available. Therefore, except for unusually small angles between the object and reference beams, the holograms recorded on photographic emulsions behave almost like truly thick holograms, and the optimum average amplitude transmittance is closer to 0.4 than 0.5. This is borne out experimentally.

For the case of thick holograms that are viewed in reflection, the theory of Chapter 7 indicates an optimum amplitude transmittance of 0, corresponding to an infinite density. In practice, good holograms are achieved at very high densities, which are easily achievable with commercially available photographic emulsions.

If $b_t(x)$ (Eq. 11.12) is constant and $\phi_t(x)$ varies, then the phase of the incident wave has been modulated and the hologram is a phase hologram. The phase of the incident wave can be modulated in one or both of two different ways. If a surface relief is formed as the result of exposure to the interference fringe pattern, then the thickness of the material is a function of x. If the index of refraction of the material is changed in accordance with the exposure, then the optical path of the transmitted light is a function of x. In either case, the phase of the transmitted light will vary with x because the optical path $n_o d$ will be a function of x. The phase ϕ_t of the transmitted light is related to the optical path according to

$$\phi_t = \left(\frac{2\pi}{\lambda}\right) n_o d. \tag{11.16}$$

Phase holograms can be recorded with several different materials; in fact the main thrust of the research that has been done in holography has been aimed at finding a good recording material for phase holograms. The best that have evolved so far are silver halide photographic emulsions, dichromated gelatin, and thermoplastic materials. Each has its own particular advantages and drawbacks, with the result that all three are used today, with the particular one depending on the application.

A phase hologram results when ϕ_t is a function of exposure (Eq. 11.14), which results when either the index of refraction n_o or the thickness d, or both, is exposure dependent. From Eq. 11.16 we see that a change in ϕ_t, denoted by $\Delta\phi_t$, is given by

$$\Delta\varphi_t = \left(\frac{2\pi}{\lambda}\right) [d\,\Delta n_o + (n_o - 1)\,\Delta d], \tag{11.17}$$

where we have written $n_o - 1$ assuming that the hologram is in air. If the hologram is thin, d is very small and the contribution to $\Delta\phi_t$ from the term $d\,\Delta n_o$ is negligible and we have

$$\Delta\varphi_t = \frac{2\pi}{\lambda}(n_o - 1)\,\Delta d. \tag{11.18}$$

This would be a so-called surface relief hologram, achieved mainly by embossing, etching a photoresist material, or with a thermoplastic film. From the theory of Chapter 7 we see that a maximum diffraction efficiency of about 0.34 is achieved when $J_1(2\gamma O_o R_o)$ is a maximum (Eq. 7.19). This occurs when

$2\gamma O_o R_o = 1.84$. From Eqs. 5.3 and 5.4 we see that $2\gamma O_o R_o$ is just the amplitude of the phase modulation, which from Eq. 11.18 is just $\frac{1}{2}(2\pi/\lambda)(n_o - 1)\,\Delta d$. (The factor of $\frac{1}{2}$ appears because Δd is usually measured peak-to-peak.) Setting this equal to 1.84 and putting $n_o = 1.5$, we find the optimum peak-to-peak relief height to be 1.17λ or approximately λ. This is rather difficult to achieve at the high spatial frequencies associated with most holograms.

For a thick hologram with a negligible surface relief, but with a varying index of refraction, Eq. 11.17 becomes

$$\Delta\varphi_t = \frac{2\pi d\,\Delta n_v}{\lambda}. \tag{11.19}$$

From Eq. 7.34 we see that a maximum efficiency of 100% is achieved when $\pi n_1 d/\lambda \cos\theta_o = \pi/2$. Since $\Delta n_o = 2n_1$, the maximum efficiency is achieved when $\Delta n_o = \lambda \cos\theta_o/d$. Taking $\lambda = 0.5\ \mu$ and a Bragg angle θ_o of 30°, we find that an index shift of about 4.3×10^{-4}/mm is required to obtain 100% efficiency for a thick phase hologram. Materials that are capable of producing this amount of index change include bleached photographic emulsions, dichromated gelatin, and ferroelectric crystals.

11.3   SILVER HALIDE PHOTOGRAPHIC EMULSIONS

Photographic films and plates were used by Gabor to record the very first holograms and are still by far the most common recording materials for holography. The longevity and widespread usage of this type of recording material can be attributed to its versatility, ease of handling, and the fact that just about every experienced scientist has used photography at one time or another in his career and so is familiar with it.

The photographic emulsion is really a dispersion of small crystals (grains) of silver bromide in a support matrix of gelatin. Sometimes silver iodide and/ or silver chloride are added in varying proportions, depending on the type of photographic material to be produced. Dyes can be added to increase the range of sensitivity from the blue and violet region of the spectrum, where the silver halide is instrinsically sensitive, to the green and red portions of the spectrum. The sensitivity or speed of the emulsion is determined primarily by the size of the silver halide grains, with the larger ones being more sensitive than the smaller ones. The reason for this is that the incident light frees electrons, which are subsequently trapped, and are thereby available for reducing a mobile silver ion. When this happens several times at the same site (as few as four silver atoms in the most sensitive emulsions), a so-called latent image speck is formed. This speck has the property that it renders the whole grain developable, no matter how large the grain is. In the presence of a strong reducing agent (the developer) the rest of the silver ions in the grain are reduced to silver atoms. These developed grains absorb light very strongly, and so a rather large optical density can be produced with the absorption of only a relatively few exposing photons. It is this tremendous gain that accounts for the very high sensitivity of photographic emulsions.

There are at present several photographic emulsions available for holography. These are manufactured by the Agfa-Gevaert Co. and the Eastman Kodak Co. and are designated as follows:

Eastman Kodak: Kodak Spectroscopic Plate, type 649F
Kodak Spectroscopic Film, type 649F
Kodak High Resolution Plate, type 649GH
Kodak Holographic Film SO-173
Kodak Holographic Plate SP-120
Kodak Holographic Film SO-253

Agfa-Gevaert: Agfa Holographic Plate 8E56
Agfa Holographic Film 8E56
Agfa Holographic Film 10E56
Agfa Holographic Plate 10E56
Agfa Holographic Film 8E75
Agfa Holographic Plate 8E75
Agfa Holographic Film 10E75
Agfa Holographic Plate 10E75

Figure 11.1 shows the spectral sensitivity curves for some of these emulsions. As indicated by these curves, the photographic emulsion is capable of recording wavelengths quite far into the red region of the spectrum. This is one of the main advantages of silver halide materials when compared to other types of recording media. This panchromatic sensitivity makes possible holographic recording with the pulsed ruby laser, one of the most important sources for holography.

The figure also shows the approximate exposure required for each of these photographic materials at each wavelength. These values are, of course, only approximate because they depend on the conditions of development and could vary by as much as a factor of 2 from the figures given for extreme processing conditions such as a very extended development time or high-temperature solutions.

These values, in ergs/cm^2, are the energy required to produce a final amplitude transmittance of 0.4, which is about optimum for an amplitude hologram viewed in transmission. For an amplitude reflection hologram the energy requirement is considerably higher because a very high optical density produces the best results. What is even more important is the case of bleached holograms, where the required exposure is 5 to 10 times greater than the

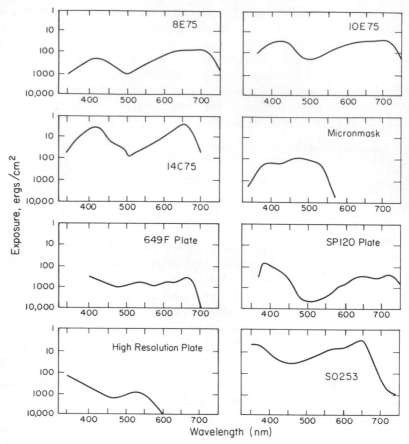

Fig. 11.1 Spectral sensitivity curves for several commercially available holographic emulsions.

values given in Fig. 11.1. This is so because in order to produce the maximum index change, the maximum amount of silver has to be formed. The 5 or 10 times figure is strictly a rule of thumb, and the exact figure depends on the particular bleach process used and on the particular emulsion. Individual experimentation should be carried out in each case.

Not much need be said about the resolution of these photographic materials since all of the emulsions listed will resolve over 1500 lines per millimeter, and this is well within the useful range for the great majority of holographic applications. In extreme cases where ultrahigh resolution is required, emulsions such as SO-173 and SP-120 have a raw (as opposed to developed) grain size of only about 450 Å, which yields a resolution capability beyond any physically realizable holographic situation.

11.4 PHOTOCHROMICS

Materials that change color reversibly when exposed to the proper optical illumination are termed photochromic. The effect has been observed in a variety of both organic and inorganic materials. Examples of inorganic photochromic materials include silver halide photochromic glasses and photochromic crystals such as KBr, CaF_2 doped with certain rare earths, $SrTiO_3$ doped with transition metals, and electron beam colored NaCl [1–5]. Examples of organic photochromic materials are photochromic spyropyran dissolved in a styrene polymer [6, 7] and a thioindigo dye [8].

Basically, the coloration of photochromic materials is caused by the presence of color centers. Color centers in transparent materials are caused by the presence of impurities and imperfections that give rise to localized states that trap electrons or holes within the forbidden energy gap of the material. These states can be absorbing at visible wavelengths, whereas the pure crystal would be transparent at these wavelengths. Photochromic materials have the ability to change colors when suitably irradiated. The change is caused by the transfer of an optically excited electron from one type of color center to another, with the absorption properties of both centers being changed. Light of one wavelength produces a given change in coloration, and the crystal returns to its original state when exposed to light of another wavelength. Typically, the photochromic material will darken upon exposure to blue or ultraviolet radiation, and then bleach out upon exposure to visible radiation. Holograms can be recorded with either the darkening or bleaching mode of operation. Figure 11.2 shows absorption spectra for a typical photochromic material.

There is usually no processing involved after exposure to produce the hologram; the hologram is formed while the exposure is being made. There is some processing involved, however, in the sense that the crystals have to be pre-processed (either darkened or bleached) by exposure to the proper radiation. Also, because of thermal bleaching, some of the crystals have to be cooled after exposure and during readout, so this can be construed as a post-exposure processing step. This is necessary, for example, with the KBr photochromic crystals.

Because of the nature of photochromism, only amplitude holograms are recorded with these materials. One of their chief advantages, however, is the fact that in their usual form (crystalline or glass) they constitute truly thick holograms; most are of the order of 1 mm thick. Other principal advantages are their ultrahigh-resolution capability (the basic mechanism is essentially molecular and so the resolution can be stated to be of the order of 10,000 cycles/mm) and their ability to be erased and reused. These features indicate a material well suited to a high-storage-capacity optical memory, and it is for this purpose that most researchers have studied various photochromics.

The main problems with photochromics are sensitivity and lifetime. Concerning the latter problem we note that inherent in the photochromic mechanism is the property that the readout light is also the erase light. This means that the readout must be either fast or done at very low illumination levels, or else some means must be found to stabilize the hologram. As mentioned previously, in some cases this is done by cooling the crystal, because even in the absence of erasing radiation, the material will revert to its normal state thermally. The maximum storage time of the crystals, which is usually determined by the thermal activities energy of the centers, varies from minutes to months.

The range of sensitivities of the inorganic photochromic materials is quite large, varying from 10^{10} to 10^5 ergs/cm². There are lasers available, principally argon, that are capable of an output in the visible portion of the spectrum sufficient to produce an exposure of this magnitude in a reasonable exposure time, but it should be borne in mind that even the most sensitive of these materials is 3 or 4 orders of magnitude slower than a silver halide emulsion. Some of the photochromic materials are actually much more sensitive in the

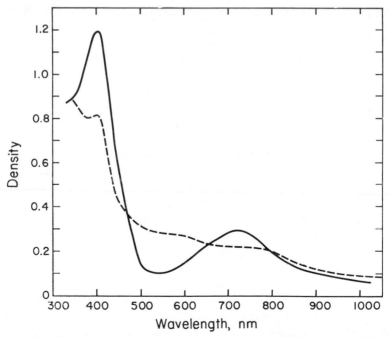

Fig. 11.2 Photochromic absorption spectra of $CaF_2:Ce$, Na in the switched (dashed) and unswitched (solid) states. (After R. C. Duncan, Jr., *R.C.A. Rev.*, **33**, 248 (1972).)

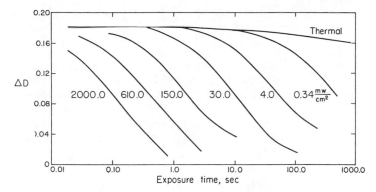

Fig. 11.3 Erase mode sensitivity of CaF_2:Ce, Na as a function of erase exposure time for various irradiances of the erase beam. The wavelength is 514.5 mm, and ΔD is the specular optical density difference between the switched and unswitched states at this wavelength. (After R. C. Duncan, Jr., *R.C.A. Rev.*, **33**, 248 (1972).)

darkening mode of operation than in the more usual bleaching mode, but this is more than offset by the unavailability of strong laser sources at the required ultraviolet wavelengths. Figure 11.3 shows some curves of the erase mode sensitivity of the same photochromic crystal as in Fig. 11.2 indicating the irradiances and exposure times typically required.

11.5 PHOTORESISTS

Photoresists are basically organic materials containing sensitizers so that a chemical and physical change occurs upon exposure to light. For a so-called positive-working resist, the exposure makes it soluble; a negative-working resist becomes insoluble upon exposure. In this mode of operation the latter are washed after exposure, so that only the insoluble portions remain, thus producing a surface relief phase hologram. It is also possible to produce volume phase holograms with photoresist materials [9]. With this mode of operation, it is not necessary to develop (wash) the resist after exposure; it is therefore a completely dry, real-time recording. The hologram is formed apparently by local changes in the index of refraction caused by the exposure.

Although many different types of resists have been tried, only a few have been found to be generally useful. The most popular seems to be the positive-working resist Shipley AZ-1350. The volume holograms made by Kurtz [9] were made on Kodak Ortho Resist (KOR), and some experiments have been done with a resist called RCA No. 1 [10].

Resist layers are prepared by coating the resist, which has been dissolved in a suitable solvent, onto a clean substrate, and then baking away the solvent.

Table 11.1

Resist	Preparation	Development
Shipley AZ-1350 (positive)	Undiluted resist spin at 5000 rpm. Bake at 75°C for 20 min.	2-min wash in Shipley developer (diluted with equal part of water). 2-min wash.
Kodak Ortho Resist (negative)	1 part resist, 1 part thinner. Spin at 800 rpm. Bake at 75°C for 10 min.	2 min bath in KOR developer. 10 sec wash with ethyl alcohol.
RCA No. 1 (negative)	Undiluted resist. Spin at 5000 rpm. Bake at 85°C for 20 min.	20 sec spray with butyl acetate. 20 sec spray with rinse (70 % isopropyl alcohol, 30 % methyl alcohol)

The photoresist can be deposited by simply dipping the substrate, but thin, uniform layers are best achieved by spinning using similar procedures to those employed in microcircuit production. Table 11.1 gives the details for preparation and development of three of the most successful resists [10]. When a negative-working resist is used, best results are obtained by using a uniform pre-exposure through the substrate to ensure hardening and thus adhesion of the underside of the resist. This ensures adhesion without the need for over-exposing the hologram.

Figure 11.4 shows the spectral sensitivities for these three resists as reported in [10]. These curves are for relatively thin films (5000 Å); it is, of course, possible to coat films much thicker than this. Also, these sensitivities are for the developed holograms. For the volume holograms produced directly, much greater exposures are required.

From these curves we see the principal problem with resists as recording materials for holography: they require a very large exposure and are sensitive only in the blue region of the spectrum. The resolution is quite high, however, and gratings have been recorded out to several thousand cycles/mm. Resists generally do very little scattering, and so produce low-noise holograms with high efficiency (diffraction efficiencies of 30 % have been reported [10].

11.6 PHOTOPOLYMERS

Photopolymers produce high-efficiency phase holograms, both by the mechanism of surface relief and by an internal change of the index of refraction [11, 12]. Basically, the material consists of a mixture of a monomer and a dye-sensitized catalyst. The holographic exposure is made at a wavelength that is absorbed by the dye in the catalyst. This produces a photopolymerization corresponding to the image, which results in both a surface relief and an internal variation of the index of refraction. The image is fixed by exposure to short wavelength radiation, which polymerizes the remainder of the monomer and converts the dye into a stable, nonabsorbing substance.

For the work of [11], the formulation was as follows:

Monomer solution:

$1.6M$ barium acrylate
$0.5M$ lead acrylate
$3.3M$ acrylamide

Photocatalyst solution:
$0.002M$ methylene blue
$0.1M$ p-toluenesulfinic acid sodium salt
$0.1M$ 4-nitrophenylacetic acid sodium salt

Two 5-cm-square cover slides were used to form a cell containing a photopolymer layer 10–20 μ thick. Exposures of 10^4–3 $\times$ 10^5 ergs/cm^2 with ruby laser radiation at 694 nm produced good holograms with high diffraction

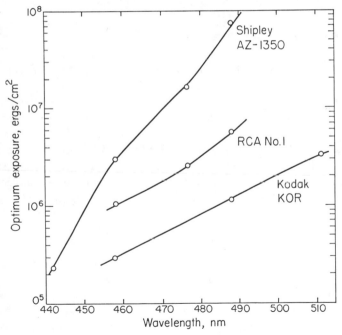

Fig. 11.4 Spectral sensitivity curves for three commercial resists.

efficiency (up to 45%). With such a high efficiency, a volume phase effect is indicated. The optical fixing procedure consisted of a 15–30-sec exposure to a 200-watt mercury arc lamp with water and Corning CS 7-54 glass filters. The holograms were reported to be quite stable following this treatment.

The work of [12], however, reports that in films of about 25 μ thickness, a surface relief mechanism is the dominant contributor to hologram formation. Enough refractive index change has been measured, however, to demonstrate that thicker holograms may be dominated by volume effects. Figure 11.5 shows a curve of refractive index versus relative exposure for this material.

DuPont has made available a photopolymer already coated onto glass plates. The polymerizable layer comprises a liquid crystal acrylic monomer, a cellulisic binder, photoinitiators, and a plasticizer. Exposure produces an index modulation and diffraction is immediate, so that the material can be used in a near-real-time mode. The uniform exposure with the reference beam also serves as the fixing beam. Extremely high diffraction efficiencies can be achieved with exposures of some 10^5 ergs/cm^2 at a wavelength of 514 nm. The sensitizing dye can be bleached out simply by exposure to a fluorescent lamp. Coburn and Haines have reported on this volume effect [13]. All of these references report that the resolution of photopolymer materials is very high, many having achieved good results up to several thousand lines/millimeter.

11.7 DICHROMATED GELATIN

It has been known for years that a layer of pure gelatin that contains a small amount of a dichromate, typically ammonium dichromate $(NH_4)_2Cr_2O_7$, is light sensitive. In fact, this is one of the earliest known photosensitive materials, dating back to the 1930s [14]. There are basically two modes of operation for dichromated gelatin (DCG), but only one is really of importance for holography. When the gelatin is soft or unhardened, DCG acts as a negative photoresist. In this mode, exposure causes the gelatin molecules to crosslink (harden) so that they become insoluble in water. The unexposed molecules, on the other hand, are unhardened and will dissolve readily in a water wash. Thus a surface relief is formed and a thin phase hologram results. The difficulty with this method of operation is that the diffraction efficiency falls off rapidly beyond a few hundred cycles/nm.

However, there is a second method that is far superior. With this method the layer is prehardened so that the entire layer becomes insoluble in water. The wash, in addition to removing the unused dichromate, produces an increase in the optical path variation above that which is produced by the exposure alone, but the diffraction efficiency is still quite low. However, a striking increase in the diffraction efficiency can be achieved by immersing

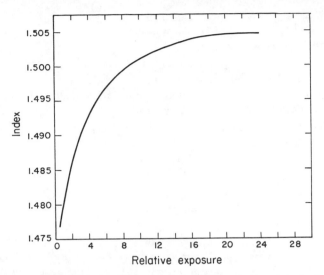

Fig. 11.5 Typical index versus exposure curve for the material of reference [12].

the layer in isopropyl alcohol following the wash. If this is followed by a very rapid drying, the optical path variation is increased to such a marked degree that diffraction efficiencies approaching the theoretical maximum of 100% can be achieved. In addition these very high efficiencies are achieved with an unusually low noise level; a properly prepared and processed DCG hologram is virtually noiseless, having the appearance of a clear glass plate.

Unfortunately, the physical mechanism responsible for this effect is not well understood. It is known that at least part of the phase change is produced by having isopropyl alcohol molecules securely bound to reduced Cr sites where the gelatin is crosslinked [15]. This effect, however, does not fully account for the observed magnitude of the index change. Curran and Shankoff have postulated that the immersion in isopropyl alcohol and the consequent rapid dehydration and shrinkage of the gelatin causes it to tear apart [16].

The basic mechanism causing the diffraction in DCG holograms is the index variation that results from the exposure and processing. Thus basically phase holograms are formed, usually thick, but in some cases approximating thin holograms if the layer has been coated very thin, but in this case the diffraction efficiency is low. One of the principal features of DCG holograms is that they can be made to appear as a piece of perfectly clear glass with virtually no scattering, and yet have a diffraction efficiency approaching 100%.

Table 11.2 gives the procedure for preparing the plates, sensitizing, and processing as described by Meyerhofer [15]. The plates can be prepared by

Table 11.2

I. Preparation of the gelatin plates.
    1. Fix 649F plates for 15 min in Kodak Rapid Fixer with Hardener.
    2. Wash for 10 min in running water at 20–25°C.
    3. Rinse with agitation for 10 min or until clear in methyl alcohol.
    4. Rinse for 10 min in clean methyl alcohol.
    5. Dry in a vertical position.

II. Preparation of sensitizer
    1. Dissolve 50 g of ammonium dichromate fine crystals in 1.0 liters of dis-
       tilled water.
    2. Filter the solution before use.

III. Sensitizing plates
    1. Soak plates for 5 min in the dichromate solution.
    2. Tilt the plate at about 10° and allow the excess to run off for about 3 min.
    3. Store in a light-tight box at a slight tilt as in (2) until ready to use.

IV. Processing
    1. Wash for 10 min in running water at 20°C.
    2. Rinse with agitation for 2 min in a 50 % mixture of isopropyl alcohol and
       water.
    3. Repeat in a 90 % isopropyl alcohol–10 % water mixture.
    4. Rinse in fresh isopropyl alcohol for 10–20 min.
    5. Pull the plate out of the alcohol very slowly (1 cm/min) while drying
       rapidly with a flow of forced hot air.

---

spreading the gelatin onto glass plates, but the method outlined here for preparing the plates from Kodak Spectroscopic Plates, type 649F is particularly convenient, although this imposes the limitation of only a single thickness.

According to Meyerhofer [15], the sensitizing solution has been chosen to incorporate the highest concentration of dichromate in the gelatin without causing undesired side effects, principally noise in the final image. The most uniform concentrations are achieved by allowing the solution to run off the nearly horizontal plates. Only about the bottom centimeter is non-uniform. The sensitized plates should be exposed between 15 and 40 hr after preparation. Using them too soon causes a crystallization, and if the plates are stored for too long, they lose sensitivity.

During processing the washed plate is immersed wet into the isopropyl alcohol. At least two steps must be used so that the last bath remains as free of water as possible.

The final step of removing the alcohol is very critical. When the plates are removed from the alcohol and simply dried in room air, the results are quite sensitive to relative humidity. Low humidity encourages crystallization while high humidity reduces the diffraction efficiency. The problem is overcome by pulling the plates slowly out of the bath and drying them with a stream of warm air directed at the liquid interface. After the drying is complete, the plates are no longer affected by humidity. The holograms are quite stable for long periods of time after processing in this way.

Dichromated gelatin plates are orthochromatic in their sensitivity. They should exhibit maximum sensitivity at around 355 nm, and then fall off to zero sensitivity at about 580 nm. According to Collier et al. [17], the relative sensitivity for the argon ion laser wavelengths of 488 and 514 nm is in the ratio of approximately 5:1. At the He-Ne wavelength of 633 nm the material is completely insensitive.

The sensitivity of DCG plates depends to some degree on the particular dichromate used, and also on the degree of hardening. Ammonium dichromate exhibits the highest sensitivity, and two beam holograms with diffraction efficiencies of 80% and greater have been achieved with exposures of 2×10^5 ergs/cm². This rather low sensitivity is one of the major drawbacks to the system. However, useful holograms can be recorded with perhaps as little as 1×10^5 ergs/cm².

The ultimate resolution of dichromated gelatin films is almost unlimited. Since it is essentially a grainless system, the resolution is limited only by the scattering of the exposing light by the gelatin and sensitizer. We can say with some assurance that the material is capable of recording any information that can physically be produced with visible light.

11.8 THERMOPLASTICS

A thermoplastic hologram is recorded as a thickness variation corresponding to the exposure pattern. They are therefore strictly thin, phase holograms. The most successful form of thermoplastic recording is the thermoplastic-photoconductor-transparent conductor sandwich operating in the charge-expose-recharge mode, as described in Section 5.3.3. Such a sandwich is shown in Fig. 11.6. Exposure causes a distribution of static charge corresponding to the image on the thermoplastic. Subsequent heating of the film softens the thermoplastic so that it deforms in accordance with the variation of the electric field corresponding to the image. The complete cycle, including erasure, is shown in Fig. 11.7. In the first step, a uniform charge is deposited on the thermoplastic with a corona charging device. The voltage, of the order of a few hundred volts, is capacitively divided between the photoconductor

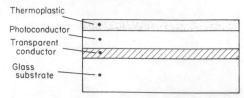

Fig. 11.6 Schematic cross section of a thermoplastic-photoconductor sandwich.

and thermoplastic layers. The second step is exposure. In the regions where the photoconductor is illuminated it conducts and so discharges the voltage across it. This occurs because in the discharge regions electrons deposit on the bottom surface of the thermoplastic. In these regions the potential at the upper surface is reduced. However the electric field in the thermoplastic, which is proportional to the surface charge density, remains unchanged. In the third step, the recharging, the potential of the upper surface of the thermoplastic is again made uniform, at a value determined by the voltage of the corona charger. Wherever the original exposure had discharged the photoconductor, the electric field increases, producing a field distribution corresponding to

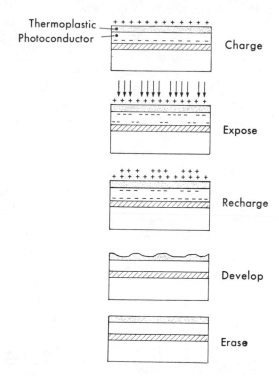

Fig. 11.7 The charge-expose-recharge cycle for thermoplastic phase holograms.

the image. In the fourth step, the sandwich is heated so that the thermoplastic softens (at a temperature between 60 and 100°C). Under the force of the varying electric field, the softened thermoplastic deforms and becomes thinner in the regions of high exposure. When the sandwich is cooled quickly to room temperature the deformation is frozen in and is stable. The device can be recycled by reheating to a higher temperature than used for development and held for a longer period of time, so that the softened material becomes unchanged and smooth.

Because of the experimental nature of the thermoplastic type of recording material, detectors of this sort are not available commercially and must therefore be prepared by the user. Basically they can be prepared as follows [17, 18]. The transparent substrate is simply a glass plate coated with a conductor such as indium oxide (available from Pittsburgh Plate Glass Co.) or tin oxide. The organic photoconductor that has been used is polyvinyl carbozole (PVK), sensitized with 2,4,7 trinitro-9-fluorene (TNF). Both of these chemicals are available from Polysciences, Inc. The Bell Laboratories workers [17] used the red sensitizing dye brilliant green. The PVK/TNF and dye are dissolved in a 1:1 mixture of 1, 4 dioxane and dichloromethane, and the substrate is pulled from the solution at a rate of between 7.5 and 13 cm/min to produce a 2.5-μ-thick coating. The proportions of solids and solvents as recommended by the Bell Labs workers [17] are PVK, 20 g; TNF, 2 g; dye, 0.2 g; solvent, 300 cc.

The thermoplastic used by these workers is Staybelite Ester 10, available from Hercules, Inc. This material is dissolved in petroleum ether, hexane, or "super hi-flash naptha" (American Mineral and Spirits Co.) [17, 18]. A 20 or 25% solution is prepared and the photoconductor coated glass is pulled through it at a rate of 5–13 cm/min. This produces a coating of 0.5–1.0 μ thickness. The complete thermoplastic-photoconductor sandwich is baked at 60°C for 1 hr. The write temperature is in the 40–50°C range and the erase temperature is approximately 70°C.

The material coated as just described is sensitive to all visible radiation, and in particular has a sensitivity of approximately 1000 ergs/cm^2 and 633 nm. This is comparable to some of the slower silver halide photographic materials, such as Kodak 649F. It is therefore quite a useful sensitivity.

A photoconductor-thermoplastic sandwich such as this has a very unusual property, at least as far as holographic materials are concerned. Because of the electrostatic nature of the recording process, the range of spatial frequencies that can be recorded is limited to a band of frequencies. Surface relief is simply not produced at the very low or very high spatial frequencies. This effect was first pointed out by Urbach and Meier [19]. The effect can be an advantage (no low-frequency noise) or a disadvantage (limited field of view). The position of the peak of the passband depends primarily on the thickness

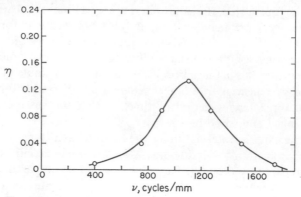

Fig. 11.8 Frequency response curve for a thermoplastic.

of the thermoplastic film and to some degree on the magnitude of the electrostatic field. The peak can occur anywhere from about 50 cycles/mm up to about 1000 cycles/mm, with an effective bandwidth of about 50%. A typical response curve for a two-beam hologram is shown in Fig. 11.8.

To summarize, we note that thermoplastic holographic recording has several very desirable features. The principal one is that it is recyclable. Recent work [20] has demonstrated that up to several thousand write-read-erase cycles are possible. Other advantages include its panchromatic sensitivity, low noise, and high diffraction efficiency. The chief disadvantages are the band-limited response and the present difficulty of preparation. The necessity of having corona chargers, heaters, and baking ovens of some sort also render processing somewhat inconvenient, although the ability to process in place for real-time interferometry is a distinct advantage.

11.9 ELECTRO-OPTIC CRYSTALS

Certain types of electro-optic crystals are suitable for holographic recording, even though their sensitivity is quite low. Exposure to light causes a distribution of the space charge density corresponding to the image in these materials. The electro-optic effect converts the resultant electric field variations into variations of the index of refraction. The index variations that are produced are quite small (of the order of 10^{-5}), and high diffraction efficiencies are therefore obtained only for fairly thick crystals. However, since the resulting hologram is a thick phase hologram, the resulting diffraction efficiencies can be quite high.

If the readout light is the same wavelength as the recording light, as is necessary for minimum aberrations, the hologram will bleach out, as with the photochromic materials. However, it has been found [21] that in lithium niobate ($LiNbO_3$) it is possible to fix the hologram by a heat treatment that cancels the internal field produced by trapped electron concentrations. A hologram fixed in this manner is optically nonerasable.

The types of electro-optic materials that have been used for holographic recording include lithium niobate ($LiNbO_3$), iron-doped lithium niobate ($LiNbO_3$:Fe), lithium tantalate ($LiTaO_3$), strontium-barium niobate [(Sr, Ba) Nb_2O_6], barium-sodium niobate ($Ba_2NaNb_5O_{15}$), barium titanate ($BaTiO_3$), and bismuth titanate ($Bi_4Ti_3O_{12}$) [22–27].

Of these, lithium niobate, and especially the iron-doped lithium niobate, have been the most successful. Diffraction efficiencies as high as 80% have been observed in 2-mm-thick iron-doped lithium niobate crystals. The main problem with these materials as holographic recording media is their insensitivity; their sensitivities lie in the range 10^7–5×10^9 ergs/cm^2 at a wavelength of 488 nm. This is some 10^6 times slower than photographic materials. Storage times up to several days are possible at ambient illumination levels (for the unfixed crystals). Most important, complete erasure of the recording can be achieved either by optical irradiation or by heating the crystal to 300°C. This means that the material is recyclable. High-quality lithium niobate crystals are readily available, which is another advantage. The main disadvantage of electro-optic materials is, of course, their insensitivity, but it is at least conceivable that in some situations this would not be too important a factor.

11.10 OTHER HOLOGRAPHIC RECORDING MATERIALS

Over the years many materials and devices have been used for recording holograms with varying degrees of success. The search for the perfect material is still going on and it may yet be found. Some of the materials and devices that are not in serious use today for various reasons include the following.

1. *Elastomer Devices.* Similar to the thermoplastic devices described above, but with the thermoplastic replaced with an elastomer. Dubbed a "Ruticon" by Sheridon [28] the device is recyclable.

2 *Magneto-Optic Materials.* Very thin (300–700 nm), semitransparent films of MnBi. A holographic exposure heats the film above the Curie temperature, switching the magnetization. Readout is through either the magneto-optic Faraday or Kerr effect. Thin phase holograms of very low diffraction efficiency are produced [29, 30].

3. *Metal Films.* A fast, high-energy holographic exposure selectively vaporizes a thin metal film to produce a thin amplitude hologram [31].

4. *Liquid Crystal Photoconductor Devices.* These consist of a layer of photoconductor material and a layer of liquid crystal sandwiched between two electrode layers. The holographic exposure results in a greater voltage across the liquid crystal, which results in increased scattering. The device has good sensitivity but low resolution [32].

5. *Ferroelectric.* Photoconductor controls the field-induced birefringence and also provides a high sensitivity. In the best mode of operation of phase hologram is produced [33, 34].

REFERENCES

[1] S. Herman, *Proceedings of the Symposium for Modern Optics*, John Wiley and Sons, New York, 1967, p. 743.

[2] J. P. Kirk, *Appl. Opt.*, **5**, 1684 (1966).

[3] R. J. Araujo, *Recent Advances in Display Media*, NASA Symposium Proceedings, Cambridge, Mass., 1967, p. 63.

[4] D. R. Bosomworth and H. J. Genitsen, *Appl. Opt.*, **7**, 95 (1968).

[5] A. S. Mackin, *Appl. Opt.*, **9**, 1658 (1970).

[6] A. L. Mikaelaine, A. P. Axenchikov, V. I. Brobinev, E. H. Gulaniane, and V. V. Shatun, *IEEE J. Quant. Elec.*, **QE-4**, 757 (1968).

[7] M. Lescinsky and M. Miller, *Opt. Comm.*, **1**, 417 (1970).

[8] D. L. Ross, *Appl. Opt.*, **10**, 571 (1971).

[9] C. N. Kurtz, *Appl. Phys. Lett.*, **14**, 59 (1969).

[10] M. I. Gale, D. L. Greenaway, and J. P. Russell, ICO International Symposium on the Applications of Holography, Besançon, France, 1970.

[11] D. A. Close, A. D. Jacobson, J. D. Marqeram, R. G. Brault, and F. J. McClung, *Appl. Phys. Letters*, **14**, 159 (1969).

[12] J. A. Jenney, *J. Opt. Soc. Am.*, **60**, 1155 (1970).

[13] W. Coburn and K. Haines, *Appl. Opt.*, **10**, 1636 (1971).

[14] J. Kosar, *Light Sensitive Systems*, John Wiley and Sons, New York, 1965.

[15] D. Meyerhoffer, *R.C.A. Rev.*, **33**, 110 (1972).

[16] R. K. Curran and T. A. Shankoff, *Appl. Opt.*, **9**, 1651 (1970).

[17] R. J. Collier, C. B. Burckhardt, and L. H. Lin, *Optical Holography*, Academic Press, New York, 1971.

[18] T. L. Credelle and F. W. Spong, *R.C.A. Rev.*, **33**, 206 (1972).

[19] J. C. Urbach and R. W. Meier, *Appl. Opt.*, **5**, 666 (1966).

[20] T. C. Lee, *Appl. Opt.*, **13**, 888 (1974).

[21] J. J. Amodei and D. L. Staebler, *Appl. Phys. Letters*, **18**, 540 (1971).

[22] F. S. Chen, J. T. La Macchia, and D. B. Fraser, *Appl. Phys. Letters*, **13**, 223 (1968).

[23] J. B. Thaxter, *Appl. Phys. Letters*, **15**, 210 (1969).

[24] J. J. Amodei, D. L. Staebler, and A. W. Stephens, *Appl. Phys. Letters*, **18**, 507 (1971).

[25] F. S. Chen, *J. Appl. Phys.*, **40**, 3389 (1969).

[26] J. J. Amodei, W. Phillips, and D. L. Staebler, *IEEE J. Quant. Elec.*, **QE-7**, 321 (1971).

[27] L. H. Lin, *Proc. IEEE*, **51**, 252 (1969).

[28] N. K. Sheridan, *J. Opt. Soc. Am.*, **60**, 723 (1970).

[29] R. S. Mezrich, *Appl. Phys. Letters*, **14**, 132 (1969).

[30] G. Fan, K. Pennington, and J. H. Greiner, *J. Appl. Phys.*, **40**, 974 (1969).

[31] J. J. Amodei and R. S. Mezrich, *Appl. Phys. Letters*, **15**, 45 (1969).

[32] H. Kiemle and U. Wolff, *J. Opt. Soc. Am.*, **60**, 1563 (1970).

[33] S. A. Keneman, G. W. Taylor, A. Miller, and W. H. Fonger, *Appl. Phys. Letters*, **17**, 173 (1970).

[34] S. A. Keneman, A. Miller, and G. W. Taylor, *Appl. Opt.*, **9**, 2279 (1970).

12 Applications for Holography

12.0 INTRODUCTION

If holography is to become a viable, useful tool in modern industry, there must be some tasks for which holography is the most suitable tool, in a practical, not fundamental, sense. The initial flurry of activity in the field after its introduction in 1948, and the vast amount of activity following the renaissance of holography in 1962, indicate that there are many people who feel that the holographic technique should prove quite useful. Indeed there have been myriad proposed uses for the technique, initially mainly concerned with microscopy, but later expanded to include interferometry, information storage, character recognition, and many others. A few applications, proposed mainly by newswriters, must be classed as science fiction. Other proposed applications, such as information storage and character recognition, represent new ways of doing old jobs, and the task of proving that the new way is better or more practical than the old is not an easy one. Finally, there are several applications of the holographic technique that do represent true innovations; these are the applications that are moving into the forefront. They represent true innovations primarily because they utilize and exploit the fundamental difference between holography and photography, namely, the fact that a hologram is a recording of a wavefront while a photograph is a recording of the irradiance distribution in an image.

Furthermore, there are different forms of holography that permit recording of information previously unattainable. These include x-ray and electron holography, and, more recently, ultrasonic holography.

As long as some of the proposed applications prove feasible, people and industry will continue to work in the field, with the result that more and more applications should be forthcoming. We will discuss in this chapter several of the most interesting proposals. Some of these do not represent unique solutions to the problem at hand, but nevertheless appear promising as a better way. Other proposed applications are unique in that there are no other means for performing the task. One of the most important in this latter category is holographic interferometry, with which we begin.

12.1 HOLOGRAPHIC INTERFEROMETRY

12.1.1 Introduction

Ever since the wave nature of light became generally accepted, interferometry has been the primary method for making measurements with great accuracy. The very small wavelength of light, of the order of 5×10^{-5} cm, and the fact that interferometric means are available for detecting changes of only a small fraction of this length, indicate the degree of accuracy that can be achieved. The widespread applications of interferometry attest to its general usefulness; it is used for testing of optical components, optical gauging of machine tools, the study of air flow in wind tunnels, and the standardization of the fundamental units of length. It is understandable, therefore, that any fundamental improvement or innovation in present interferometric technique would find many applications.

Holographic interferometry is such a fundamental innovation. Holography has widened the scope of interferometry to such a degree that it is hard to believe that holographic interferometry will not be used as a standard tool in engineering laboratories all over the world in a few years.

Conventional interferometry can be used to make measurements on highly polished surfaces of relatively simple shape. Holographic interferometry extends this range by allowing measurements to be made on three-dimensional surfaces of arbitrary shape and surface condition. A roughly machined machine part can now be measured to optical tolerance. Furthermore, with the holographic technique, a complex object can be examined interferometrically from many different perspectives, by means of the three-dimensional nature of the hologram. A single interferometric hologram is equivalent to many observations with a conventional interferometer. This property is especially useful for observations of such things as fluid flow in a wind tunnel [1]. A third departure of holographic interferometry from conventional interferometry is that an object can be interferometrically examined at two different times; one can detect with wavelength accuracy any changes undergone by an object over a period of time. The present object can be compared with itself as it was at an earlier time. This is a very great advantage in many fields. For example, a large lens can be tested before and after mounting. With the use of pulsed lasers, a machine part can be interferometrically compared with itself statically and dynamically.

One could mention many other specific applications for this powerful new technique, but we will instead discuss in more detail the major proposed applications and methods of holographic interferometry. Specifically these are the single- and double-exposure holographic interferogram, vibration analysis, contour generation, and pulsed laser interferometry.

12.1.2 Single-Exposure Holographic Interferometry

To understand the basic process of holographic interferometry, first consider a hologram of an object wave $O(x)$ recorded as shown in Fig. 12.1. The object wave at the hologram plane is denoted by $O(x)$. This can be written in the usual form

$$O(x) = O_o(x)e^{i\varphi_o(x)} \tag{12.1}$$

where $O_o(x)$ is the (real) amplitude distribution and $\varphi_o(x)$ is the phase distribution in H. The reference wave at the hologram is written

$$R(x) = R_o e^{ikx \sin \alpha_R} \tag{12.2}$$

where R_o is the amplitude of the reference wave at the hologram and α_R is the off-axis angle. We assume that the exposure and resulting amplitude transmittance is simply $|H(x)|^2$, where $H(x) = O(x) + R(x)$; therefore

$$|H(x)|^2 = O_o^2(x) + R_o^2 + R_o O_o(x)[e^{i[\varphi_o(x) - kx \sin \alpha_R]} + e^{-i[\varphi_c(x) - kx \sin \alpha_R]}]. \tag{12.3}$$

Next assume that we replace the hologram in exactly the same position it occupied during exposure and illuminate it with the wave

$$C(x) = O'(x) + R(x). \tag{12.4}$$

The wave $O'(x)$ means that we are illuminating the hologram with the object still in place, but we are allowing for a slight change in the object between recording the hologram and illuminating it. The wave $O'(x)$ is similar to $O(x)$, and we write it as

$$O'(x) = O_o(x)e^{i\varphi_o'(x)}, \tag{12.5}$$

that is, it has the same amplitude distribution but a slightly different phase

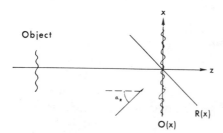

Object

Fig. 12.1 Recording the object wave in the standard manner.

distribution. The wave transmitted by the hologram becomes

$$\psi(x) = C(x) \cdot |H(x)|^2$$
$$= O_o^3 e^{i\varphi_o'} + O_o R_o^2 e^{i\varphi_o'} + R_o O_o^2 e^{i(\varphi_o - \varphi_o' - kx\sin\alpha_R)}$$
$$+ R_o O_o^2 e^{-i(\varphi_o - \varphi_o' - kx\sin\alpha_R)} + R_o O_o^2 e^{ikx\sin\alpha_R}$$
$$+ R_o^3 e^{ikx\sin\alpha_R} + R_o^2 O_o e^{i\varphi_o} + R_o^2 O_o e^{-i(\varphi_o - 2kx\sin\alpha_R)}. \quad (12.6)$$

This transmitted field consists of many waves traveling in several directions; the directions are determined by the linear variations of x in the exponents. The first, second, and seventh terms on the right-hand side of Eq. 12.6 represent component waves traveling in the approximate direction of the original object wave. These terms represent the new object wave $O_o(x)e^{i\varphi_o'(x)}$, which is transmitted directly through the hologram (first and second terms) and the reconstructed object wave (seventh term). Considering these terms alone, we have a transmitted component

$$\psi'(x) = O_o(x)\{[O_o^2(x) + R_o^2]e^{i\varphi_o'(x)} + R_o^2 e^{i\varphi_o(x)}\}. \quad (12.7)$$

The irradiance that the eye or camera sees is given by

$$|\psi'(x)|^2 = O_o^2(x)\{[O_o^2 + R_o^2]^2 + R_o^4 + 2R_o^2[O_o^2 + R_o^2]\cos(\varphi_o' - \varphi_o)\}. \quad (12.8)$$

The irradiance distribution consists of a series of fringes; the location of a fringe is defined by the locus of all points for which

$$\varphi_o'(x) - \varphi_o(x) = \text{const.} \quad (12.9)$$

A dark fringe is produced whenever

$$\varphi_o'(x) - \varphi_o(x) = (2m + 1)\frac{\pi}{2}, \qquad |m| = 0, 1, 2, \ldots. \quad (12.10)$$

Figure 12.2 indicates the two interfering wavefronts. To show that $\psi'(x)$ is

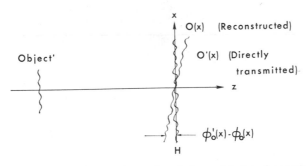

Fig. 12.2 The two interfering wavefronts in a holographic interferometer. The object, labeled OBJECT′ in the figure, yields a wave $O'(x)$ at the hologram. This wave interferes with the reconstructed wave $O(x)$.

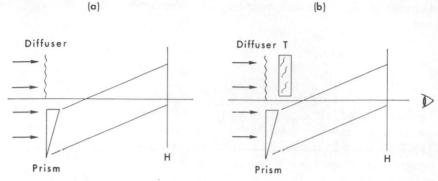

Fig. 12.3 A general-purpose holographic interferometer. In (*a*), a hologram of a diffuser is recorded. In (*b*), the hologram is placed exactly in its original position so that both the reconstructed and actual object wave are superposed. When the test plate *T* is inserted, any optical inhomogeneities will distort the actual object wave so that interference fringes appear.

approximately the original object wave, we write $e^{i\varphi_o'(x)} = e^{i[\varphi_o(x)+\delta(x)]}$ so that Eq. 12.7 becomes

$$\psi'(x) = O_o(x)e^{i\varphi_o(x)}[(O_o^2 + R_o^2)e^{i\delta(x)} + R_o^2] \qquad (12.11)$$

where $\delta(x)$ is the deviation of the changed object wavefront from the original. In most cases, $O_o^2(x)$ will be the rapid transmission variations resulting from the speckle pattern at the hologram from a diffuse object. Thus the wave $\psi'(x)$ just appears to be the original object wave modulated by transmission through this speckle pattern. Since $O_o(x)$ is already a rapidly varying function of x, the imposition of $O_o^2(x)$ on $O_o(x)$ does not yield a visible difference in the appearance of the image. If $\delta(x)$ is a smoothly and slowly varying function of x, the transmitted wave $\psi'(x)$ yields an image of the object crossed with fringes.

Another useful form of the single-exposure hologram interferometer is shown in Fig. 12.3. In (*a*), a hologram of a diffusing plate is recorded. After processing, the hologram is repositioned accurately. When the hologram is illuminated with the original reference beam, with the diffuser still in position, the primary image falls exactly in the position still occupied by the diffuser itself. The system is adjusted so that these two waves lie precisely on top of each other and a single bright fringe occupies the field of view. If now a test flat *T* (Fig. 12.3*b*) is inserted between the diffuser and hologram, but not intercepting the reference beam, fringes appear in the image that are related to the optical inhomogeneities, wedge, and thickness variations present in the test place. The operation of this sort of interferometer is similar to the common Twyman-Green interferometer, except that the optical quality of *T* may be examined from many perspectives. Figure 12.4 shows a typical result that can be obtained using this type of interferometer. The test plate T is a microscope cover glass. The two photos show the resultant interference pattern for two perspectives.

The single-exposure holographic interferometer may also be used to measure surface deformation in arbitrary objects at two different times [2]. A hologram of an object is made (Fig. 12.5a). The hologram is accurately replaced in its original position and the object is stressed. In (b), the hologram is illuminated with the reference beam and the light scattered from the stressed object. A view of the object viewed through the hologram is shown in (c). Interference fringes are present that are related to the degree of deformation of the object surface, and quantitative calculations can be made of the exact amount of deformation present, although not simply. Figure 12.6 is an interferogram of this type, with the object under two amounts of strain. The stress was introduced in the object by tightening the C-clamp slightly. This same technique can be used to measure surface deformations of all kinds, including thermal expansion and contractions, swelling due to absorption, and any minute changes that occur in any object.

The single-exposure hologram interferometer is particularly useful for observing phenomena within a transparent object, regardless of the optical quality of the envelope material. A hologram is made with the object placed behind the diffusing plate of Fig. 12.3a. Then, if the object changes in any way, fringes will appear within the object. Suppose the object were a gas-filled lamp, for example. A hologram is made of the object with the lamp off. If the lamp is now turned on, so that the gas heats up, fringes will be observed due to the changes in optical path through the gas (see Fig. 12.9, below). The

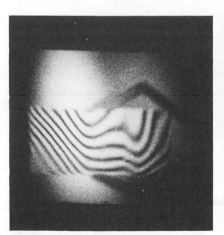 

Fig. 12.4 Two views of an interferogram of a microscope cover slide made with the interferometer of Fig. 12.3.

optical quality of the envelope is immaterial, since the object is being compared with itself. The optical path variations due to imperfections in the envelope are automatically canceled. This would not be true for a conventional interferometer, where the object wave would be compared with a constant reference, such as a plane wave. In this case, fringes would result from the interference of the wave deformed by the envelope and the reference, resulting in an extremely complicated fringe pattern in which it would be

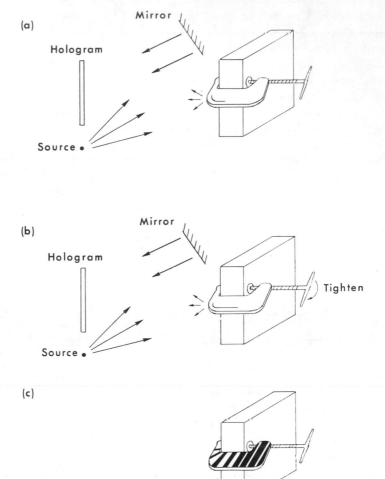

Fig. 12.5 Holographic strain interferometry. In (a), a hologram is recorded of an object. In (b), the hologram is replaced in its exact original position and the object is strained. In (c), light from the strained object passes directly through the hologram and interferes with the reconstructed wave of the unstrained object.

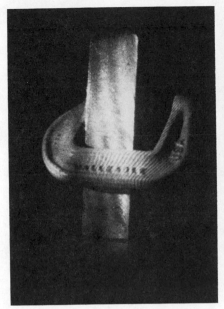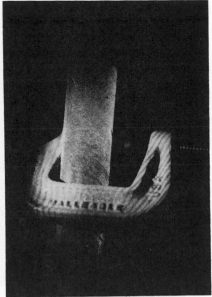

Fig. 12.6 Two different amounts of stress on a C-clamp, recorded as described by Fig. 12.5.

difficult to distinguish between the permanent variations of the object and the variations induced by heating. For this reason, holographic interferometry has been called differential interferometry, detecting only changes in an object.

12.1.3 Double-Exposure Holographic Interferometry

The double-exposure holographic interferogram is similar to the single-exposure case in most respects. However, accurate registration of object and hologram is no longer required. Precision optical components are not required and it is still a differential interferometer; in addition, a complete three-dimensional record of the interference phenomena is obtained, permitting postexposure focusing and examination from various directions, just as in the single-exposure case.

The reason for the differences between single- and double-exposure holographic interferometry is that with the double-exposure method the hologram retains as a permanent record the change in shape of the object between exposures. The two interfering waves may be reconstructed with a separate arrangement, without the need for an accurate registration of the plate in its original position. By using this technique, a permanent record can be made of differential changes in an object over a period of time.

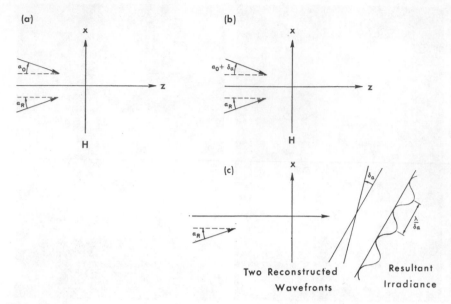

Fig. 12.7 A simple example of a double-exposure interferogram. In (a) a hologram of a plane wavefront is recorded. In (b), the hologram is doubly exposed by recording another hologram of a plane wavefront incident at a slightly different angle. In (c), both of these waves are reconstructed and they interfere so as to form straight fringes of spacing $\lambda/\delta\alpha$.

The double-exposure interferogram requires making two holograms on a single recording medium. One of the two holograms yields a primary image that constitutes the comparison wave, just as in the single-exposure case. However, the test wave is not the object itself, but a reconstructed wave from the changed object. Interference phenomena, caused by changes in optical path through the object between exposures, are produced when the double hologram is illuminated.

As a simple example of the double-exposure method, suppose we make a hologram of a plane wavefront, propagating at an angle α_o to the axis (Fig. 12.7a). The reference wave, also plane, is traveling at an angle α_R to the axis. The irradiance at the hologram plane H is

$$|H|^2 = |e^{iax} + e^{ibx}|^2 = e^{i(a-b)x} + e^{-i(a-b)x} + 2 \qquad (12.12)$$

where $a = k \sin \alpha_o$ and $b = k \sin \alpha_R$; the amplitudes of the object and reference waves are taken as unity. Now suppose that a second exposure is made on the same recording medium, only that now the object beam is incident at a slightly different angle, $\alpha_o + \delta\alpha$ (Fig. 12.7b). This hologram exposure is

$$|H'|^2 = |e^{i(a+\epsilon)x} + e^{ibx}|^2$$
$$= e^{i(a+\epsilon-b)x} + e^{-i(a+\epsilon-b)x} + 2 \qquad (12.13)$$

where $\epsilon \cong k(\delta\alpha) \cos \alpha_o$ for $\delta\alpha$ small. If we now reconstruct by illuminating with the reference wave e^{ibx}, the transmitted wave becomes

$$\psi(x) = 4e^{ibx} + e^{iax} + e^{-i(a-2b)x} + e^{i(a+\epsilon)x} + e^{-i(a+\epsilon-2b)x}. \qquad (12.14)$$

The second and fourth terms on the right-hand side represent the two reconstructed primary wavefronts (Fig. 12.7c):

$$\psi_p(x) = e^{iax} + e^{i(a+\epsilon)x}. \qquad (12.15)$$

A detector placed behind the hologram will detect an irradiance proportional to

$$|\psi_p(x)|^2 = 2 + 2 \cos \epsilon x, \qquad (12.16)$$

so that there are fringes in the transmitted wave representing the displacement of the object wave. The fringe spacing along the tilted wavefront is given by

$$\Delta(x \cos \alpha_o) = \frac{\lambda}{\delta\alpha}. \qquad (12.17)$$

Another way of interpreting the fringes is as a spatial beat frequency. Each hologram consists of a single spatial frequency exposed onto the recording medium, the second spatial frequency being a little larger than the first. The composite hologram thus consists of a fringe pattern that is periodically washed out and reinforced. In the region of the washed-our fringe pattern, the wavefront does not reconstruct, which gives rise to the modulated wavefront on the exit side of the hologram. One way to observe this spatial beat frequency would be to incoherently illuminate the hologram, and then look at the hologram itself. Dark bands, representing the spatial beats, or moire pattern, are then observable in the regions where the grating pattern was washed out by path differences of an odd number of half wavelengths.

The operation of the double-exposure hologram interferometer with diffuse objects is analogous to the single-exposure case. Conceptually, one can extend the preceding ideas for the plane wave case to the diffuse object wave case by considering the object wave as a set of plane waves propagating in various directions. The composite hologram will again contain the moire beats between the two hologram exposures, resulting in the fringes on the reconstructed waves.

The double-exposure technique is well suited to interferometric recording of transient phenomena, such as shock waves and fluid flow, when a pulsed laser is used as the source. All of the above principles apply equally well to time-dependent events, and the very short pulse of light from a ruby laser can record the interference phenomena at a single instant of time. The very wide range of applicability of the method has been well demonstrated by Brooks, Wuerker, Heflinger, and their co-workers at TRW, Inc. [3–6]. Two excellent examples of their work are shown in Figs. 12.8 and 12.9.

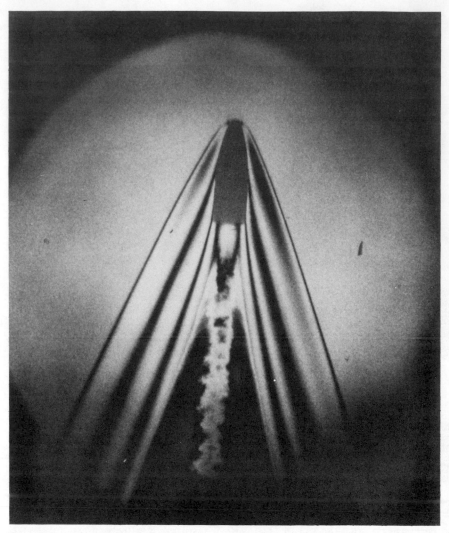

Fig. 12.8 A double-exposure holographic interferogram of the shock wave of a 0.22 caliber bullet traveling at 3500 ft/sec. One exposure was made before the bullet was fired and the second just as the bullet and shock wave passes into view. The hologram was recorded using a Q-switched ruby laser and illuminated with the light from a He-Ne laser. A diffuser behind the object permitted it to be viewed from different angles, although the photo is taken at only one point of view. (After Brooks, Wuerker, Heflinger and Knox [6]; courtesy of TRW, Inc.)

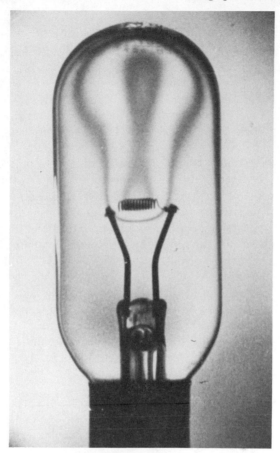

Fig. 12.9 The double-exposure holographic interferogram as a differential interferometer. One exposure was made with the filament cold and the second with the filament heated. The difference between the two objects is only the density change of the filling gas. The interference pattern describing this change is accurately displayed interferometrically in spite of the poor optical quality of the envelope material. (After Brooks, Wuerker, Heflinger and Knox [6]; courtesy of TRW, Inc.)

Although we have only discussed double exposures in this section, it is known that this technique can be extended to multiple exposures, with some interesting results [7]. Suppose, for example, that the object changes by some small increment between each exposure. The holographic image will contain fringes corresponding to the change of shape of the object, but the bright fringes will be much sharper than in the double-exposure case. Multiple-exposure holographic interferometry, as with ordinary multiple-beam interferometry, yields sharp fringes, so that fringe displacement may be measured with great accuracy.

12.1.4   Fringe Interpretation

The problem of relating the observed fringes to the actual subject deformation is complex and has been treated by many authors [8–14]. In an extensive series of papers, Abramson [15–19] has demonstrated that in most cases interpretation of the fringes need not be difficult. He has introduced a diagram that he has called a "holo-diagram." The holo-diagram is a diagram that allows measurements of deformation even for relatively complex and large objects. The diagram shown in Figure 12.10 enables the engineer to determine the sensitivity to motion of the object at each point of the object. The diagram is assumed to be superposed on the holographic table, with point A representing the source point of the object and reference beams and B a point on the hologram. These two points become the foci of ellipses drawn in the plane of the diagram. The point C represents a point of the object. Light emanating from point A illuminates the object at C and is then scattered to the point B of the hologram. The path length for the light rays ACB is constant for all points C on the ellipse having foci A and B. This is the significance of the ellipses. The path length of the light changes, however, if the point C moves off the ellipse. It is this component of the motion that gives rise to the observed fringes.

The reference beam is derived in any convenient manner, provided only that the path length is approximately equal to the object beam path. An easy

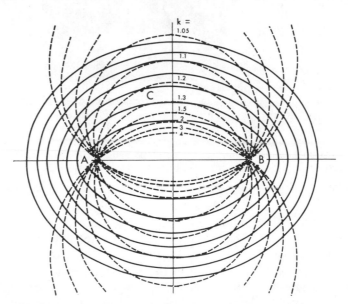

Fig. 12.10   The holo-diagram [15].

way to ensure this is to place a mirror tangent to the same ellipse that C is on. Then there will be exactly zero path difference at point B of the hologram.

In any real situation the object will not lie simply in the plane of the diagram, and so it must be remembered that in reality the ellipses of the plane diagram are ellipsoids with foci A and B. In almost all cases this will not present a problem because the object motions of interest will be in a direction that is more or less perpendicular to the ellipses and in the ABC plane.

Because the ellipses are not equally spaced everywhere on the diagram, the number of fringes generated by a given object motion will depend on the position of C in the diagram. The most sensitive position for C is on the x-axis outside of A and B. Clearly, every time the object point C moves along this line a distance of $\lambda/2$, the path length ABC changes by λ and so the illumination at B changes one full cycle, say from light to dark to light again. The least sensitive position of the diagram is on the x-axis between A and B. Here motion along the axis does not affect the path length ABC. At any given position in the diagram the greatest number of cycles corresponding to a given movement is produced when the movement is perpendicular to the ellipses. The nearer to the center of the diagram, the greater the movement that is required to give one cycle of illumination variation at B. The sensitivity of the hologram is determined by the number of half wavelengths of object displacement required to cause a full cycle of illumination variation at B. Call this number k. The quantity k is thus a measure of the sensitivity, and the magnitude of k is a function of position in the diagram. At any point in the diagram an object motion of $k\lambda/2$ produces one full cycle of illumination (one fringe) at the point B of the hologram. The value of k is one along the x-axis to the right of B and along the y-axis at an infinite distance from AB. The value of k increases towards the center of the diagram and in the center it is infinite. At all other positions k is between 1 and ∞. The locus of all points on the diagram having constant k-value form arcs of circles, which are shown in the diagram with the value of k indicated. The value of k can be quickly and easily determined through the relationship

$$k = \sec \frac{\alpha}{2}$$

where α is the angle ACB. By knowing k, in many cases it is an easy task to determine the amount an object has moved simply by counting fringes, either as they move by in a real-time situation, or from a fixed point of the object to the point C in a double-exposure hologram. Even if there is no fixed point on the object, there are ways to determine the object displacement [15].

The techniques for fringe interpretation presented by Abramson, and other methods, for example, the method of Hecht et al. [14], have gone a long way toward making holographic interferometry a routine method of measurement.

As more people become familiar with the techniques of holographic interferometry, its use will become more and more universal; even now it has been tried and used in almost every industrial laboratory the world over.

12.1.5 Time-Average Holographic Interferometry

The idea of multiple-exposure interferometry may be extended to the limiting case of a continuum of exposures, resulting in what might be called time-average holographic interferometry. This technique lends itself well to the problem of vibration analysis, and may be the best method yet devised for such analysis. The earliest report of this form of holographic interferometry was by Powell and Stetson [20]. The basic idea of the method is that since holography itself is an interferometric process, any instabilities of the interferometer cause fringe motion. Thus the hologram of a vibrating object is a record of the time-averaged irradiance distribution at the hologram plane. Since the amount of light flux diffracted from any region of the hologram depends on the fringe contrast, any object motion that causes the fringes to move during the exposure, causing a loss in contrast, will result in less diffracted flux from that region of the hologram. The strength of the reconstructed wave is therefore a function of the fringe motion during the exposure. One can think of the process as a recording of very many incremental holograms, one for each incremental position of the object. In the reconstruction, each incremental hologram yields an image wave, each slightly displaced, producing the interference effects.

If the object is vibrating in a normal mode, there will be standing waves of vibration on the surface, so that at the nodes the object motion will be very small or nonexistent. At the antinodes, the vibration amplitude will be large. A hologram of such an object will then reproduce a bright image of the regions of the object for which little or no motion occurred during the exposure, while not at all reproducing the antinodal points. Figure 12.11 shows some of the exceptional photos taken by Powell and Stetson using this technique. The vibrating surface is the end of a cylindrical can. The nodes are the bright areas of the photo, and by counting the contours lying between the nodes and a known stationary object point, the amplitude of vibration for each object point can be found, since each bright area represents an amplitude of vibration of approximately a multiple of a half wavelength, depending on the recording geometry.

This holographic method of vibration analysis has all of the advantages of all holographic interferometry. The method can be used regardless of the shape or complexity of the object; the vibration nodes can be examined in three dimensions, or at least from a variety of perspectives; and the method works regardless of whether the surface is optically smooth or diffusely reflecting. The technique should find widespread application.

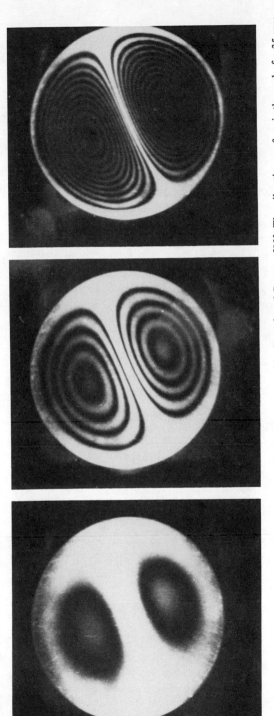

Fig. 12.11 Some of the time-average holographic interferograms produced by Powel and Stetson [20]. The vibrating surface is the end of a 35 mm film can.

235

12.1.6   Contour Generation

The idea of contour generation was first proposed by Haines and Hildebrand [21] and should find some interesting and useful applications. The basic idea behind the process can be easily explained. Suppose, as shown in Fig. 12.12a, a hologram is made of a single point object a distance z_o from the hologram plane H. The reference wave is a plane wave making some angle with the hologram plane. Let us further assume that the light source used for this recording contains two wavelengths, λ_1 and λ_2. Since the light at these two wavelengths has no mutual coherence, the recorded hologram is really two incoherently superposed holograms of the point object. We now illuminate the hologram with light of wavelength λ_1 only. The illuminating wave is also plane, and is incident on the hologram at the same angle as the reference wave, so that a spherical wave is reconstructed that appears to have originated at the point at z_o. This reconstruction is produced by the component hologram recorded with the light of wavelength λ_1. However, the other component hologram, recorded with the light of wavelength λ_2, reconstructs another spherical wavefront, with a different radius (Fig. 12.11b). This is because of the magnification resulting from recording and reconstructing with different wavelengths. From Eq. 9.132, the expression for the radius of a reconstructed spherical wavefront is

$$\mathbf{Z}_p = \frac{z_c z_o z_R}{z_o z_R + \mu z_c z_R - \mu z_c z_o} \tag{12.18}$$

where the hologram scaling factor m is unity and the wavelength ratio μ is the ratio of the wavelengths used for recording and illumination. In this case, $z_c = z_R = \infty$ so Eq. 12.18 becomes

$$\mathbf{Z}_p = \frac{z_o}{\mu}. \tag{12.19}$$

(a)   (b)

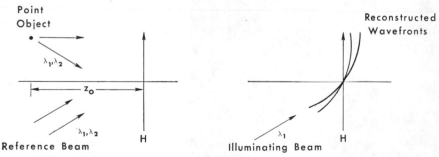

Fig. 12.12   Holographic contour generation. (a) Recording an object point with two wavelengths. (b) Reconstructing two wavefronts with different magnifications.

Now for the hologram recorded at λ_1, $\mu = 1$, and for the hologram recorded at λ_2, $\mu = \lambda_1/\lambda_2$. Hence the two spherical waves produced have radii z_o and z_o/μ, so that a viewer looking through the hologram sees two virtual images, one at a distance z_o from the hologram and the other at a distance z_o/μ. The optical path difference between these two wavefronts is

$$OPD = \Delta z_o = \left(\frac{z_o}{\mu}\right) - z_o = \frac{z_o}{\mu}(1 - \mu) = z_o \frac{\Delta\lambda}{\lambda_1} \qquad (12.20)$$

where $\Delta\lambda = \lambda_2 - \lambda_1$. For most cases, $\Delta\lambda$ is chosen small enough so that the actual image observed appears to be a single point. However, whether or not the point is visible at all depends on how far the point is from the hologram. For values of z_o such that

$$\Delta z_o = m\lambda_1, \qquad |m| = 0, 1, 2, \ldots \qquad (12.21)$$

a bright point will be observed. However, for values of z_o such that

$$\Delta z_o = (m + \tfrac{1}{2})\lambda_1 \qquad (12.22)$$

the point will not be observable, because of destructive interference between the two waves. Passing to the continuous case, we see that for a continuous array of points in an object, only those portions of the object at a distance z_o from the hologram will appear in the image, where z_o is given by the condition

$$z_o \frac{\Delta\lambda}{\lambda_1} = m\lambda_1, \qquad |m| = 0, 1, 2, \ldots. \qquad (12.23)$$

Therefore, the image of an object recorded with two discrete wavelengths separated by $\Delta\lambda$ will appear to have fringes on the surface, with each fringe being the locus of all points at a constant distance from the hologram. The fringes are contours of constant depth of the object, with the change in depth between fringes given by

$$\delta z_o = \frac{\lambda_1^2}{\Delta\lambda}, \qquad (12.24)$$

which follows immediately from Eq. 12.23. Clearly a more rigorous analysis would include consideration of the viewing angle, since the optical path difference between the two waves is a function of position in the hologram plane.

The occurrence of $\lambda^2/\Delta\lambda$ in Eq. 12.24, which is just the coherence length of the light used to make the hologram, indicates an interesting use of this technique. Suppose the object to be very long and positioned as in Fig. 12.13. If a hologram is made of this object, only those points along the object satisfying Eq. 12.23 will yield bright images—for other values of z_o, the corresponding image points will be less intense—and the intensity distribution in the image will be directly related to the coherence properties of the source.

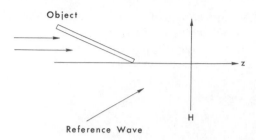

Object

z

H

Reference Wave

Fig. 12.13 A possible holographic arrangement for mapping the fringe visibility function of a light source.

If the source is a single line of width $\Delta\lambda$, the image brightness will decrease from a maximum to zero over a distance $\delta z_o = \lambda^2/\Delta\lambda$. If the source contains two discrete lines separated by $\Delta\lambda$, the image will be fringed as discussed above, since the fringe visibility function, as given just following Eq. 10.27, is periodic with period $\lambda^2/\Delta\lambda$. For more complicated distributions, the intensity distribution in the image is more complicated, but is still proportional to the visibility function. Therefore, we see that this technique may also be used as an interferometric spectrometer.

12.2   CORRECTING ABERRATED WAVEFRONTS WITH HOLOGRAMS [22]

The use of extra optical devices in imaging systems to improve their performance is not uncommon; the most common image improving device is the stop. Use has been made of Fresnel zone plates, phase plates, and other devices for the purpose of improving optical images. Holograms, too, may be used for this purpose, but their use has several drawbacks that severely limit the usefulness of this technique. Nevertheless, some special cases may arise wherein the hologram technique could prove quite useful. Also, knowledge of the process of aberration correction with holograms is useful for anyone attempting to make holograms with good image resolution where lenses are used in making the hologram and forming the image, since it is possible to compensate at least partially for the lens aberrations with proper technique. The basic system used for making the hologram is shown in Fig. 12.14*a*. Suppose the lens L produces a wavefront W with aberrations. The reference wave R is plane. The wave W in the hologram plane is described by

$$O(x) = e^{i\varphi_0(x)} \tag{12.25}$$

and the reference wave by

$$R(x) = e^{ikx\sin\alpha_R} \tag{12.26}$$

as usual, where we are assuming unit amplitude. After processing, the amplitude transmission of the hologram is assumed to be simply

$$|H|^2 = |R + O|^2 = 2 + e^{i[\varphi_o(x) - kx \sin \alpha_R]} + e^{-i[\varphi_o(x) - kx \sin \alpha_R]}. \quad (12.27)$$

Now since the lens L has been used to collimate the light, the function $\varphi_o(x)$ is just the phase difference between W and an ideal plane wavefront. However, the last term on the right-hand side of Eq. 12.27 contains a phase term that is conjugate to W; hence the hologram can be used as a corrector plate.

To do this, the hologram is illuminated from behind so that a wavefront W^* is produced (Fig. 12.14b). That is, the hologram is illuminated in such a way that all of the rays of Fig. 12.14a are exactly reversed. The lens L should then produce a diffraction limited point image in its focal plane, since the incident wavefront W^* is expressed as $e^{-\phi_o(x)}$, which will just be canceled when multiplied by the lens phase error, $e^{i\phi_o(x)}$.

Figure 12.15 shows the results achieved by Upatnieks et al. [22] with this method. These are the point images produced by a poor lens (a) without the hologram corrector plate, (b) with the hologram corrector plate, and (c) a $25 \times$ enlargement of (b). Note that this last is almost an Airy disk.

Although the analysis above considered only a collimated input to the lens, there is also improvement over a finite field, as evidenced in the photographs of Fig. 12.16. In (a) the lens is used without the hologram corrector plate, while in (b) the corrector plate is used. The improvement of the image quality over the field is obvious.

The drawbacks to the method include the fact that little light is diffracted by the hologram into the desired order; only 2–3 % is typical. However, it should be possible to make phase holograms with more diffraction efficiency, which would overcome this problem. Another disadvantage of the method is that monochromatic light must be used because of the dispersion of the hologram,

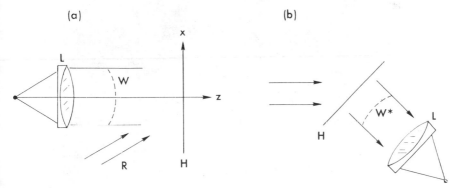

Fig. 12.14 Holographic correction of lens aberrations. (a) Recording the aberration (b) The hologram produces a wave that is phase conjugate to the aberrated wave. When the lens L images this wave, the errors will cancel.

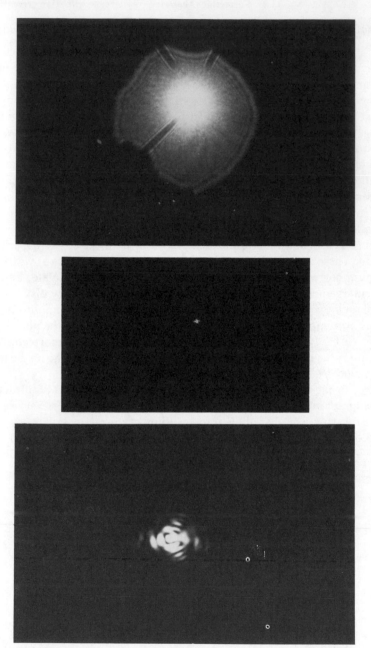

Fig. 12.15 Actual examples of holographic correction. (*a*) The point spread function of a lens with a large amount of spherical aberration. (*b*) The spread function of the lens when used with the holographic corrector plate. (*c*) A 25× enlargement of (*b*).

Fig. 12.16 Holographic correction over an extended field. (*a*) Image using lens without corrector plate. (*b*) Image using lens with corrector plate.

which limits the general usefulness of the method still more. Methods are available for achromatization, such as use of prisms or gratings, but the system rapidly becomes quite complex. Finally, the main drawback is that the hologram corrector plate introduces new aberrations of its own. As shown in Chapter 9, whenever the illuminating wave is not identical to the reference wave, aberrations are introduced. If the corrector plate has been made for use with a certain lens, and the lens subsequently used to image a scene, not all points of the scene will illuminate the hologram from the same direction. Thus aberrations will be added to the waves from these scene points. The analysis of [22] shows how these additional aberrations may be minimized but never eliminated. The authors show that, under optimum conditions, there will still be some residual astigmatism and distortion introduced by the hologram corrector plate. Hence the scheme of correcting lenses with holograms only corrects some aberrations at the expense of increasing others.

12.3 MATCHED FILTERING AND CHARACTER RECOGNITION

The use of holograms as optical spatial filters was introduced by Vander Lugt [23] almost simultaneously with the introduction of holograms made with off-axis reference beams. Basically, the optical matched filter is a photographic recording of the Fourier transform of the amplitude distribution of an object. Since it is the amplitude distribution with which we are concerned, it is necessary to record the phase as well as the irradiance in the Fourier transform, and this is why holograms are so well suited to this problem.

The important problem of character recognition is one for which holograms might be used advantageously, since optical character recognition is one of the prime uses for the optical matched filter. How the optical matched filter is used for character recognition can be described simply as follows: A number of masks are stored on a master hologram. Each mask corresponds to one of the characters to be read. An optical filtering process performs a cross-correlation between an unknown character and the master hologram. If the mask corresponding to the unknown character in the input plane is present, a bright spot appears in the output plane in a position that corresponds to that character.

The master hologram consists of a large number of Fourier transform holograms of a set of characters to be stored. It is produced as shown in Fig. 12.17. The field distribution in the x-y plane is the Fourier transform of the set of characters (Appendix A). The reference wave is plane and is denoted by R in Fig. 12.17.

The completed hologram is replaced as indicated in Fig. 12.18. When an unknown character is placed in the window in the x-y plane, the lens L produces its Fourier transform at the hologram. Here the Fourier transform of

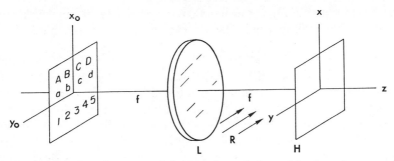

Fig. 12.17 Producing the master hologram for a holographic character recognition system.

the unknown character is multiplied with all of the recorded Fourier transforms (by the process of transmission through the hologram). A second lens L produces a Fourier transform of this product in the x_i-y_i plane, and therefore the cross-correlations of the unknown character and all of the stored characters are displayed in this plane. Since the cross-correlation of two functions is the integral of their product as a function of relative displacement, a bright spot will appear in the x_i-y_i plane when the unknown character is correlated with a stored character of similar shape. This bright central spot corresponds to the integral over the product at zero displacement. This central bright spot will be surrounded by flare light, corresponding to the cross-correlations of the character to be read with all of the other stored masks. An array of detectors in the x_i-y_i plane would then determine the position of the brightest spot, which in turn identifies the unknown character.

The process may be described mathematically as follows: Suppose we make the mask of a single character $C(x_o)$. The total amplitude in the hologram plane (the x-y plane of Fig. 12.16) is then (in one dimension)

$$H(x) = R(x) + O(x) = R_o e^{ikx\sin\alpha_R} + \left(\frac{-i}{\lambda f}\right)^{1/2} \int_{-\infty}^{\infty} C(x_o) e^{i(k/f)xx_o} dx_o \quad (12.28)$$

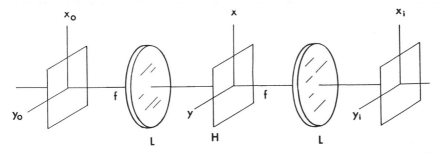

Fig. 12.18 Identifying an unknown character in the input plane with a holographic character recognition system.

where R_o is the amplitude of the plane reference wave, which makes an angle α_R with the hologram normal. The recorded irradiance in this plane is

$$|H(x)|^2 = R_o^2 + |O(x)|^2 + R_o O(x)e^{-ikx\sin\alpha_R} + R_o O^*(x)e^{ikx\sin\alpha_R}. \quad (12.29)$$

It is the last term on the right-hand side which is of interest to us here, so we retain only that term. This term would normally lead to the conjugate image. Now let us suppose that the amplitude transmission of the processed hologram is simply equal to $|H(x)|^2$. The hologram is replaced in the x-y plane (Fig. 12.17). If now the character $C(x_o)$ is placed in the aperture of the system (x_o-y_o plane), a field

$$O(x) = \left(\frac{-i}{\lambda f}\right)^{1/2} \int_{-\infty}^{\infty} C(x_o')e^{i(k/f)xx_o'}\,dx_o' \quad (12.30)$$

is incident on the hologram. The transmitted wave is then given by

$$\begin{aligned}\psi(x) &= O(x)O^*(x)R_o e^{ikx\sin\alpha} \\ &= \frac{R_o}{\lambda f}e^{ikx\sin\alpha_R}\int_{-\infty}^{\infty}C(x_o')e^{-i(k/f)xx_o'}\,dx_o'\int_{-\infty}^{\infty}C(x_o)e^{-i(k/f)xx_o}\,dx_o. \quad (12.31)\end{aligned}$$

Combining these integrals yields

$$\psi(x) = \frac{R_0}{\lambda f}e^{ikx\sin\alpha_R}\iint\limits_{-\infty}^{\infty}C(x_o)C(x_o')e^{i(k/f)x(x_o'-x_o)}\,dx_o'\,dx_o. \quad (12.32)$$

The second lens L of Fig. 12.17 now forms the image in the x_i plane:

$$G(x_i) = \left(\frac{-i}{\lambda f}\right)^{1/2}\int_{-\infty}^{\infty}\psi(x)e^{i(k/f)xx_i}\,dx. \quad (12.33)$$

Substituting Eq. 12.32 into 12.33 gives

$$G(x_i) = \frac{R_o}{\lambda f}\left(\frac{-i}{\lambda f}\right)^{1/2}\int_{-\infty}^{\infty}\left[e^{ikx\sin\alpha_R}\iint\limits_{-\infty}^{\infty}C(x_o)C(x_o')e^{i(k/f)x(x_o'-x_o)}\,dx_o'\,dx_o\right]e^{i(k/f)xx_o}\,dx_i.$$

$$(12.34)$$

By interchanging the order of integration we get

$$G(x_i) = \frac{R_o}{\lambda f}\left(\frac{-i}{\lambda f}\right)^{1/2}\iiint\limits_{-\infty}^{\infty}C(x_o)C(x_o')e^{i(k/f)x(x_o'-x_o+x_i+f\sin\alpha_R)}\,dx_o\,dx_o'\,dx \quad (12.35)$$

Now by writing

$$\delta(a) = \frac{1}{2\pi}\int_{-\infty}^{\infty}e^{-iax}\,dx, \quad (12.36)$$

and integrating over x, we obtain

$$G(x_i) = \frac{R_o}{\lambda f}\left(\frac{-i}{\lambda f}\right)^{1/2} 2\pi \iint\limits_{-\infty}^{\infty} C(x_o)C(x_o') \, \delta\left[\frac{k}{f}(x_o' - x_o + x_i + f\sin\alpha_R)\right] dx_o \, dx_o'$$

$$= R_o\left(\frac{-i}{\lambda f}\right)^{1/2} \int_{-\infty}^{\infty} C(x_o)C(x_o - x_i - f\sin\alpha_R) \, dx_o. \tag{12.37}$$

This is just the desired correlation function. It is represented by a bright spot in the x_i plane at the position $x_i = -f\sin\alpha_R$. To show that this position is a function of the position of the character during recording of the master hologram, we assume that the character was originally centered at a point a so that it is represented by $C(x_o - a)$ and

$$O(x) = \left(\frac{-i}{\lambda f}\right)^{1/2} \int_{-\infty}^{\infty} C(x_o - a)e^{i(k/f)xx_o} \, dx_o. \tag{12.38}$$

The term of interest in the hologram exposure thus becomes

$$R(x)O^*(x) = R_o e^{ikx\sin\alpha_R}\left(\frac{-i}{\lambda f}\right)^{1/2} \int_{-\infty}^{\infty} C(x_o - a)e^{-i(k/f)xx_o} \, dx_o. \tag{12.39}$$

Now when this same character is placed in the center of the aperture, the illumination of the hologram is

$$O(x) = \left(\frac{-i}{\lambda f}\right)^{1/2} \int_{-\infty}^{\infty} C(x_o')e^{i(k/f)xx_o'} \, dx_o' \tag{12.40}$$

and the transmitted wave becomes

$$\psi(x) = \frac{R_o}{\lambda f} e^{ikx\sin\alpha_R}\iint\limits_{-\infty}^{\infty} C(x_o - a)e^{-i(k/f)xx_o}C(x_o')e^{i(k/f)xx_o'} \, dx_o \, dx_o'. \tag{12.41}$$

Therefore the image is given by

$$G(x_i) = \left(\frac{-i}{\lambda f}\right)^{1/2} \int_{-\infty}^{\infty} \psi(x)e^{i(k/f)xx_i} \, dx$$

$$= \frac{R_o}{\lambda f}\left(\frac{-i}{\lambda f}\right)^{1/2} \iiint\limits_{-\infty}^{\infty} C(x_o - a)C(x_o')e^{-i(k/f)x(x_o - x_o' - x_i - f\sin\alpha_R)} \, dx_o \, dx_o' \, dx$$

$$= \frac{R_o}{\lambda f}\left(\frac{-i}{\lambda f}\right)^{1/2} \cdot 2\pi \iint\limits_{-\infty}^{\infty} C(x_o - a)C(x_o') \, \delta\left[\frac{k}{f}(x_o - x_o' - x_i - f\sin\alpha_R)\right] dx_o \, dx_o'$$

$$= R_o\left(\frac{-i}{\lambda f}\right)^{1/2} \int_{-\infty}^{\infty} C(x_o - a)C(x_o - x_i - f\sin\alpha_R) \, dx_o$$

$$= R_o\left(\frac{-i}{\lambda f}\right)^{1/2} \int_{-\infty}^{\infty} C(y)C(y + a - x_i - f\sin\alpha_R) \, dy, \tag{12.42}$$

which has its maximum value at the point

$$x_i = a - f \sin \alpha_R. \tag{12.43}$$

Hence the bright spot representing the character originally described by $C(x_o)$ was located at $x_i = -f \sin \alpha_R$, while the spot representing the character $C(x_o - a)$ is located at $x_i = a - f \sin \alpha_R$. Thus the presence of a bright point in the output plane indicates the presence of a character in the aperture, and its position corresponds to the original position of the character when the master hologram was made, which identifies the character.

Burckhardt [24] has considered the problem of how many masks of individual characters can be stored on a single hologram. For a completely noiseless recording medium, he finds for the maximum number N of masks that can be stored

$$N = \frac{\pi v_c^2 D^2}{32 a^2 v^2} \tag{12.44}$$

where

  v_c = highest recordable spatial frequency
  D = diameter of lens L in Fig. 12.16
  α = dimension of input character
  v = maximum spatial frequency in the spectrum of the input character

These N characters are stored on the master hologram with P separate exposures, where

$$P = \frac{N}{M_{\max}} \tag{12.45}$$

and $M_{\max}$ is the greatest number of characters that may be stored in a single exposure:

$$M_{\max} = \frac{\pi D^4}{256 f^2 \lambda^2 v^2 a^2} \tag{12.46}$$

where f is the focal length of the lens L in Fig. 12.17, and λ is the wavelength of the recording light. $M_{\max}$ is a finite number because of the finite resolution of the lens L. Assuming the parameters

$$f = 5 \times 10^2 \text{ mm} \qquad\qquad D = 10^2 \text{ mm}$$
$$\lambda = 6.33 \times 10^{-4} \text{ mm} \qquad a = 3 \text{ mm}$$
$$v_c = 1.6 \times 10^3 \text{ cycles/mm} \quad v = 10 \text{ cycles/mm}$$

we find that $N \approx 2.5 \times 10^6$ and $M_{\max} \approx 10^4$.

For the case of a nonperfect recording medium, the maximum number of characters that may be stored is somewhat less for a given signal-to-noise ratio. The general equation in this case is given by

$$N = \frac{(T_o - 1)^2 F_1}{256(S/N)^2 R \cdot P_1} \cdot \frac{D^2}{a^2 \nu^2},$$ (12.47)

where T_o is the average amplitude transmission of the master hologram, F_1 is a constant ($\sim$30) that takes into account the fact that the area occupied by the correlation function (12.37) is not uniformly filled with light, S/N is the desired signal-to-noise power ratio, and R is the fraction of the incident power scattered per (unit bandwidth)^2. Taking $T_o = 0.5$, $F_1 = 30$, $S/N = 100$, $R = 6 \times 10^{-9}$, and $P = 1$, we find

$$N \approx 5 \times 10^3.$$ (12.48)

This is, of course, smaller than that formed for the noiseless system, but is still sufficiently large to store all of the alpha-numeric characters, each with some 140 variants.

12.4 HOLOGRAPHIC MICROSCOPY

Holography and microscopy are linked together historically. Holography was invented by Gabor in 1948 in an effort to improve the imagery of the electron microscope. In those early years other workers [25, 26] attempted to extend Gabor's ideas to include x-ray microscopy. Although these early attempts to apply the ideas of holography to microscopy never really fulfilled the expectations of the experimenters, current work indicates that success may yet be achieved in the areas of x-ray and electron holographic microscopy, and other forms of holographic microscopy have already proven quite successful. The reason for the current interest is basically the same as that for the renewed interest in holography itself: the improvements to Gabor's early methods introduced by Leith and Upatnieks in 1962. These workers also did some of the first successful microscopy using the new methods [27]. Later workers [28–30] have shown that a different technique can be employed that allows all of the conventional types of illumination: bright field, dark field, phase contrast, interference contrast, and polarization microscopy.

The main problem with holographic microscopy is that magnification is achieved by means of a wavelength and/or radius of curvature change between recording and reconstructing. These changes are always accompanied by the wavefront aberrations discussed in Section 9.4. The elimination and reduction of these aberrations was discussed there, and one has to design the holographic microscope so as to eliminate most aberrations.

Generally speaking, holographic image resolution is limited by the diffraction limit of the hologram aperture, assuming, of course, that such things as bandwidth, source size, and aberrations are all made negligible, as discussed in Chapter 9. Therefore, resolution in microscopy depends mainly on the wavelength used to make the hologram. This is why electron waves and x-rays appear so interesting.

The image magnification, on the other hand, depends on such things as the radii of curvature of the wavefronts used to record and illuminate the hologram, the ratio of the wavelengths used in recording and reconstructing, and the linear magnification of the hologram itself. The general formula for magnification was derived in Chapter 9:

$$M = \frac{m}{1 \pm m^2 z_o/\mu z_c - z_o/z_R}, \tag{12.49}$$

where the upper sign is for the primary image (magnification as viewed from the hologram of this image is virtual) and the lower sign for the conjugate image. The linear magnification of the hologram is denoted by m and the wavelength ratio by μ. As indicated in Chapter 9, any magnification caused by an increase of the illuminating wavelength must be accompanied by a corresponding scaling up of the hologram; otherwise image resolution will be lost by an increase in the aberrations.

Equation 12.49 explains the interest in recording the hologram with x-rays or electron waves and reconstructing with light waves. Since the wavelength of x-rays or electron waves is so short, μ can become quite large, resulting in large magnifications. The expression also indicates how magnification can be achieved by changing the relationship between the geometrical positions (z_o, z_R, z_c) of the object, reference, and reconstructing beams. Gabor [31], El Sum [25], and Baez [26] changed the wavelength to achieve high magnification. Leith and Upatnieks [27] and Thompson et al. [32] changed the geometry.

There are several features of the wavefront-reconstruction technique that make it potentially more suitable (at least in some areas) for microscopy than conventional imaging techniques. One of these features is the fact that in holography the field of view is a function of the resolution and size of the recording material. We can therefore expect, at least theoretically, to obtain good imagery over much larger fields than is attainable with conventional microscopy at high magnifications. Further, since the complete wavefront has been reconstructed, the hologram contains information about the object through a much greater depth than is possible with conventional microscopy. Still another advantage of holographic microscopy is that the image can be examined at some time after the hologram has been recorded. Thus when a time-varying phenomenon has to be studied, holographic microscopy can be employed to examine the object as it appeared at one instant. Furthermore, changes can be artificially induced in the object and the changed specimen can

be compared with itself before it was changed. Since the hologram reconstructs the actual object wave, the various image-processing techniques that
are normally applied to the original object can still be applied. This is a feature
unique to holography.

There are basically two distinct methods of holographic microscopy: (1)
conventional holography with magnification achieved by changing the scale
of the hologram, the illuminating wavelength, or the radius of curvature of the
illuminating wavefront; optically magnifying the holographic image; or using
any combination of these; or (2) holographically recording an optically
magnified wavefront. We begin with a discussion of the first method.

The main problem associated with this method is that magnification is
usually achieved by means of a wavelength and/or radius-of-curvature change
between recording and reconstructing. (Few attempts have been made at
scaling the hologram.) These changes are always accompanied by the usual
wavefront aberrations shown in Tables 9.1–9.3. The holographic microscope
must be designed so as to eliminate most of these aberrations. Any magnification caused by increasing the illuminating wavelength must be accompanied
by a corresponding scaling up of the hologram; otherwise, resolution will be
lost by an increase in the aberrations.

One of the earliest and most successful applications of holographic microscopy was particle size determination [32]. Conventional methods of
physically sampling a dynamic aersol are unsatisfactory because they are too
slow. By recording a hologram of the aerosol using a pulsed ruby laser, one
can examine a three-dimensional volume of moving particles in detail at a
later time. The useful recorded depth of field is very great compared to that of
a conventional photomicrograph.

To illustrate the advantage of the holographic method, consider the problem of trying to record by means of conventional microscopy the image of two
particles, 10 μ in diameter, that are separated longitudinally by 1 cm. It is
impossible to build a conventional system that will record both particles
simultaneously. A microscope that can resolve the 10-μ particles would have a
depth of field of the order of 100 μ, not 1 cm. This problem can be solved by
holography, which is especially useful in dynamic situations when it is not
possible to focus and record each member of a sample of particles separately.

Figure 12.19 is a schematic diagram of a fog hologram camera used by

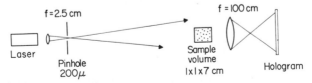

Fig. 12.19 Schematic of the fog hologram camera [32].

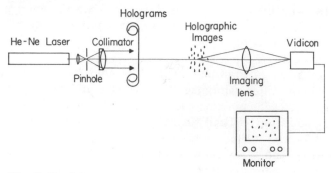

Fig. 12.20   Schematic of the readout system [32].

Thompson et al [32]. There is no separate reference beam since this is a Gabor type of hologram. The sample volume recorded on each hologram is 7 cm³ and the system is capable of 6-μ resolution. Illumination is by a pulsed ruby laser with a pulse duration of 0.5 μsec. Figure 12.20 is a schematic of the readout system. A He-Ne laser and collimator provide a plane wave to illuminate the hologram and form an image of the volume of particles. The over-all magnification is about 300×.

Knox [33] has also demonstrated the usefulness of this technique. He used a Q-switched ruby laser, capable of producing coherent light pulses of 100-nsec duration, to record living marine plankton organisms. He thus recorded, at a single instant of time, a complex, transient, microscopic event, throughout a significant volume for later analysis with conventional microscopic equipment. The total volume that he recorded, using the illumination system shown in Fig. 12.21, was over 1000 cm³. Using a conventional microscope to view a real image formed with the hologram allowed any point in the volume to be brought into sharp focus.

Figure 12.22*a* is a schematic diagram of a holographic microscope such as used by Leith and Upatnieks [27]. The specimen *O* is placed a distance z_o from the hologram *H*. For reconstruction (Fig. 12.22*b*), the wavelength may be different from that used in recording, and the hologram (denoted by *H'*) scaled accordingly. The illuminating source is located a distance z from *H'*,

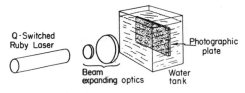

Fig. 12.21   Schematic of the illumination system used by Knox to record instantaneously a large volume of marine plankton organisms [33].

(a)   (b)

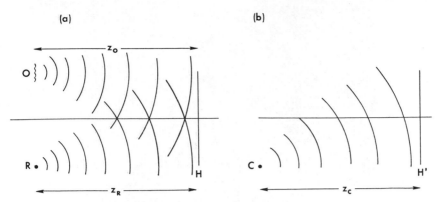

Fig. 12.22 Schematic of a holographic microscope. (*a*) A hologram of the object *O* is recorded with a spherical reference wave derived from a point source a distance Z_R from the hologram plane. (*b*) The hologram, which may have been scaled, is denoted by H' and is illuminated with a spherical wave derived from a point source a distance Z_c from the hologram.

which is generally different than the distance z_R of the reference source from the hologram plane during recording.

    Figures 12.23*a* and *b* show two examples of what has been achieved in practice [27]. For these, $\lambda_r = \lambda_c = 0.633\ \mu$ so that $\mu = m = 1$. Figure 12.23*a* is a portion of a fly's wing taken at $M = 60$. Figure 12.23*b* is a bar pattern taken at $M = 120$. The spacing between the closest bars is about 8 μ. Van Ligten [34] reports that the achievable resolution is even less for general objects, a result that he attributes to problems with the photographic emulsion and substrate.

    Figure 12.24 is a schematic diagram of the holographic microscope used by van Ligten and Osterberg [28] to record holographically a magnified wavefront. This is essentially identical with the system used by Ellis [29], who first demonstrated the technique. An unmodified reconstruction yielded an image that was a faithful reproduction of the original microscope image with bright-field illumination. Using this *same hologram*, Ellis was able to obtain:

1. Dark-field reconstruction. The image is formed by blocking the direct light at the aperture.

2. Spatial filtering. Fine image detail was removed by reducing the aperture.

3. Bright-ground positive phase contrast. A lightly darkened mask with a clear central aperture was used as a mask.

4. Dark-ground negative phase contrast. A lightly darkened spot attenuated and retarded the direct light.

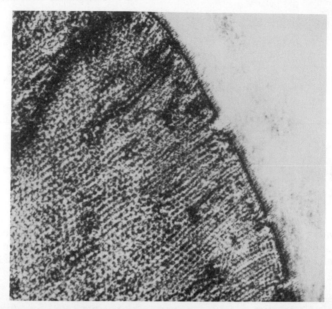

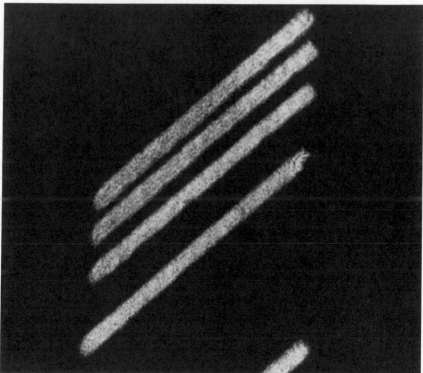

Fig. 12.23 Examples of holographic microscopy, often Leith and Upatnieks [27]. (*a*) Fly's wing, $m = 60$, $\lambda = 0.633\,\mu$, $m = \mu = 1$. (*b*) Bar pattern, $m = 120$, $\lambda = 0.633\,\mu$ $m = \mu = 1$. The spacing between the closest bars is about $7\,\mu$.

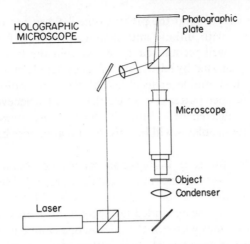

HOLOGRAPHIC
MICROSCOPE

Photographic
plate

Microscope

Object
Condenser

Laser

Fig. 12.24 Schematic of the holographic microscope used by van Ligten and Osterberg [28].

5.   Interference. A small lenticular wedge divided the beam into two parts to produce shear fringes.

Polarization microscopy has been demonstrated by van Ligten [34]. The hologram is made with a laser emitting linearly polarized light. Normally, the interference at the hologram plane during recording takes place between the components of the reference and object beams that have the same polarization. This means that the hologram contains information of the specimen as if it were taken between a polarizer and an analyzer set parallel. Introducing a rotator in one of the beams [35] will allow reconstruction of the specimen representing information that would be obtained with the polarizer and analyzer set at an arbitrary angle with each other. This is done during recording and poses no difficulties during reconstruction.

By way of summary, we should note that holography can circumvent the problems of limited depth of focus, off-axis aberration, and the relatively short working distance of the classical microscope. Theoretically, the hologram records a large volume of object space without loss of resolution. Also, since the hologram contains all the information about the object, all the usual image-processing techniques can be applied to the holographically reconstructed wave. These include bright-field illumination, dark-field illumination, spatial filtering, bright-ground positive phase contrast, dark-ground negative phase contrast, interference microscopy, and polarization microscopy. Holographic microscopy also allows standard microscopic techniques to be applied to dynamic objects throughout a large volume in object space. This feature is unique to the holographic method.

Holographic microscopy at the present time is faced not only with serious problems, but also with unique and exciting prospects for the future. Its principal difficulty, well recognized by workers in the field, consists of the diffraction noise generated by edges, out-of-focus details in the specimen, dust particles in the system, and laser speckle. This last is the noise associated with the random interference patterns in space that occur whenever coherent light is transmitted or reflected in a diffuse fashion. Most investigators believe, however, that these problems will be solved, and advances have already been made.

Prospects for the future in holographic microscopy seem to indicate that conventional microscopy can be fundamentally augmented by the capabilities of the coherent-light microscope. The types of spatial filtering of possible use to the microscopist may be expanded far beyond those presently employed. It has been shown that many uses of the microscope can be highly simplified or automated in the performance of such routine tasks as counting and sizing blood cells or other assemblies of small objects [36].

12.5   IMAGERY THROUGH DIFFUSING MEDIA [37]

One possible use for holograms might be detection of a spatial signal in the presence of spatial noise, either stationary or time varying. As an example of the former, consider the simple case illustrated in Fig. 12.25. The object transparency, described by a transmission function $t(x)$, is invisible as viewed from the hologram plane H, because of the presence of the diffusing plate described by the transmission function $d(x)$. However, it is possible to form an image of the transparency alone by using the conjugate image. To do this, one forms a hologram in the usual way, as indicated in the figure. We suppose that the disturbance at the hologram plane, in one dimension, can be

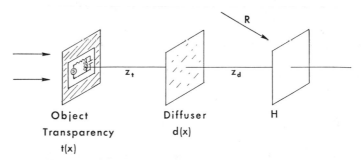

Object   Diffuser   H
Transparency   d(x)
t(x)

Fig. 12.25   Recording the hologram of an object obscured by a diffusing plate $d(x)$.

represented by a compound function

$$O(x) = T(x) + D(x) \tag{12.50}$$

where $T(x)$ is the portion of the total object wave due to the transparency and $D(x)$ is the diffuser component, which makes observation of $T(x)$ alone impossible. Addition of a reference wave $R(x)$ yields an exposure (omitting constants)

$$
\begin{aligned}
|H|^2 &= |T + D + R|^2 \\
&= |T + D|^2 + |R|^2 + R^*(T + D) + R(T^* + D^*).
\end{aligned}
\tag{12.51}
$$

By illuminating the hologram with a wave $R(x)$ (Fig. 12.26), one of the reconstructed waves (conjugate wave) is

$$R^2(T^* + D^*). \tag{12.52}$$

By the time this wave has propagated a distance z_d (the distance of the diffuser from the hologram plane during recording), $D^*(x)$ has become $d^*(x)$, that is, the complex conjugate of the original diffuser function $d(x)$. If the identical diffuser $d(x)$ is placed in this position, the wave transmitted by the diffuser becomes $d(x)\,d^*(x) = d^2 = $ const., provided that $d(x)$ is pure imaginary, that is, a phase disturbance only. Thus the noisy part of the reconstructed wavefront has been filtered out; only the portion corresponding to the transparency remains. By the time the wave (12.52) has propagated a distance z_t, the function $t^*(x)$ is imaged, easily recognizable as the object transparency.

This procedure presents an interesting coding scheme. A recognizable image will result only if the exact diffuser $d(x)$ is available, and it must be relocated in exactly the correct position. If $d(x)$ is not available, $t(x)$ can never be isolated. If the image (conjugate) of the diffuser and the diffuser itself are not precisely coincident, a noisy pattern instead of $t^*(x)$ is produced.

If the noisy medium between the object and hologram is time varying, and thus irreplaceable, the preceding method is of no help. However, there is still a holographic technique that yields a better image than could be obtained

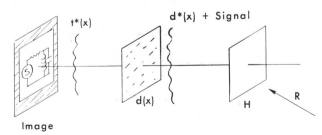

Fig. 12.26   Recovering an image of the originally obscured object by using the diffuser as a complex spatial filter.

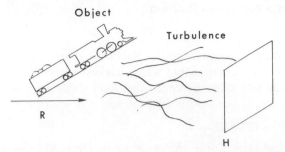

Object

Turbulence

R

H

Fig. 12.27   Recording a hologram of an object obscured by a turbulent atmosphere. By keeping the reference beam close to the object beam, the phase difference between the two waves at the recording plane is substantially independent of the turbulence; the hologram is recorded almost as if there was no turbulence.

conventionally [38]. Suppose, as indicated in Fig. 12.27, an object is viewed from some plane H, but the time-varying, low spatial frequency phase perturbations of the intervening medium tend to obscure the image. As examples, atmospheric turbulence and rough water would represent such media. If a hologram of the object is recorded such that the reference wave traverses substantially the same path as the object wave, it is possible that a usable image can be formed. This is possible, because, to a first approximation, the random phase variations introduced onto the object wave are also imposed on the reference wave. Therefore, since the phase difference between these two waves at the hologram will be largely independent of the intervening medium, the resulting hologram will be substantially the same, regardless of the presence or absence of the perturbing medium.

12.6   THREE-DIMENSIONAL OBSERVATION

One of the most striking aspects of the modern hologram is the three-dimensional image that it is capable of producing. This three-dimensional image indicates that there is a large amount of information contained in a single hologram, certainly much more than is contained in a conventional photograph of the same size. This is especially true when the object in question is many times the depth of field in depth. Because of the many perspectives which are available, the hologram is well suited to display purposes. With a hologram, one can present all of the observable characteristics of a three-dimensional object in a clear and concise manner. Complicated molecular or anatomical structure can be simply presented with a single holographic image, with little chance of error or misinterpretation on the part of the viewer. Such a hologram would take the place of several conventional drawings or photographs. The use of holograms in textbooks would be a great aid to the student in many areas. Holograms made so that they are viewable with a small pen light and a colored filter have already been produced in large quantities and distributed in magazines and books.

The use of holographic images for simulation systems has some drawbacks, such as magnification and/or power for illumination, but these are difficulties easily overcome if the need and usefulness outweigh the costs. Their use as training devices may well prove to be quite advantageous.

12.7 INFORMATION STORAGE

12.7.1 Introduction

The use of holograms as information storage devices has been one of the most promising right from the start. Initially one tends to think that a hologram is capable of storing much more information on a two-dimensional medium than a photograph. This is certainly true if the photograph is merely an image of the information. However, there are more suitable ways to store information photographically that make the contest between holography and photography about even. The question of whether holography is the best means for storing information has tantalized scientists and engineers. The question has not yet been resolved, but there is still significant effort being applied in this area.

Further, when one adds the third dimension—the depth of the holographic recording medium—it becomes clear that holography looks very promising indeed. The gain in storage capacity by utilizing the depth of a photographic emulsion is very real, but not nearly as significant as the gains that may be realized with the use of the very thick (of the order of 1 mm or more) photochromic materials.

The proven information capacity of a two-dimensional photographic emulsion is of the order of 10^8 bits/cm², whether the information is in the form of a binary code or microimage. This is for a signal-to-noise (S/N) ratio of about 10. This is the number holography must surpass. On a commercial basis, conventional imaging techniques are used for storage capacities of the order of 10^3 bits/cm² with a S/N of the order of 10^3. For holographic data storage, in two dimensions, the capacities are about the same. However, holographic data storage techniques offer two possible advantages: the possibility of the utilization of a three-dimensional recording medium and a large redundancy because of the way in which the data are stored.

12.7.2 Coherent and Incoherent Superposition

Before we make an order-of-magnitude estimate of the storage capacity of a hologram, we would discuss the method of recording. In conventional optical

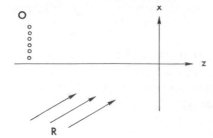

Fig. 12.28 A hypothetical hologram recording system for determining the advantages of coherent versus incoherent superposition of holograms.

data storage, the information is stored in the form of either a binary code (small patches of density or no density, or possibly several density levels), with the presence or absence of density representing one bit of information, or in the form of microimages of the information. With holographic data storage, each bit is stored as a single frequency grating pattern. Each point in object space thus represents a bit of information, and this information is stored over an extended area of the holographic recording medium. With the hologram method, each bit can be recorded separately through multiple exposures (incoherent superposition) or, alternatively, all of the bits can be recorded simultaneously (coherent superposition). The information stored in a hologram is read out by reconstructing the plane (or spherical) object wave corresponding to each bit. The strength of the readout signal is proportional to the variance of the amplitude transmittance of the hologram, and this is proportional to the input exposure modulation. We can show that for coherent superposition, this modulation decreases as $N^{-1/2}$ with increasing N, while for incoherent superposition it decreases as N^{-1}, where N is the number of bits recorded. This means that more flux will be diffracted per bit of information when the bits are recorded coherently than when they are recorded incoherently.

To show this, we refer to Fig. 12.28. The object O consists of many discrete points, each representing one bit of information. For the case of coherent superposition, all object points are illuminated simultaneously, so that the light from each object point interferes with the reference beam at the hologram. Also, each object point will interfere with each other point.

For incoherent superposition, each object point interferes with the reference beam sequentially. The reference beam is now attenuated so that the ratio of irradiance at the hologram from a given object point to that of the reference beam is the same as the ratios of irradiance of the total object to reference beam in the case of coherent superposition, so that the total exposure for each case is the same. Since the object points are exposed in sequence, the interference pattern due to an object point and the reference beam will simply add in irradiance to the other interference patterns.

We first assume that the amplitude at the hologram from each of the N object points can be described by $O_m e^{i\phi_m}$, and that all of the object waves have the same amplitude O_1. The reference wave is described by $R_o e^{ikx}$. The total irradiance at the hologram plane due to the object is NO_1^2, and the beam balance ratio is

$$K = \frac{NO_1^2}{R_o^2}. \tag{12.53}$$

When the hologram is recorded, the exposure is

$$E = \left| O_1 \sum_{n=1}^{N} e^{i\varphi_m} + R_o e^{ikx} \right|^2$$
$$= NO_1^2 + R_o^2 + 2O_1 R_o \cos [\varphi_1 - kx] + \cdots. \tag{12.54}$$

This contains a large number of cross terms that appear as cosine functions having various arguments. However, we are only interested in knowing the modulation of only one such component. This modulation is given by

$$M_c = \frac{2O_1 R_o}{NO_1^2 + R_o^2} \tag{12.55}$$

where the subscript c refers to the modulation for coherent superposition. If we let $A_t^2 = NO_1^2$ be the total irradiance from the object, we have

$$M_c = \frac{2A_t R_o}{N^{1/2}(A_t^2 + R_o^2)}. \tag{12.56}$$

For the case of incoherent superposition, the light from a single object point interferes with the reference beam. The exposure in this case is

$$E_1 = |O_1 e^{i\varphi_1} + N^{1/2} R_o e^{ikx}|^2$$
$$= O_1^2 + \frac{R_o^2}{N} + 2 \frac{O_1 R_o}{N^{1/2}} \cos [\varphi_1 - kx]. \tag{12.57}$$

The total exposure received will result from the superposition of N such interference patterns,

$$E = \sum_{n=1}^{N} \left\{ O_n^2 + \left(\frac{R_o^2}{N} \right)_n + 2 \frac{O_1 R_o}{N^{1/2}} \cos [\varphi_n - kx] \right\}$$
$$= NO_1^2 + R_o^2 + 2 \frac{O_1 R_o}{N^{1/2}} \sum_{n=1}^{N} \cos [\varphi_n - kx]. \tag{12.58}$$

The beam balance ratio for each exposure of the form (12.57) is

$$K = \frac{O_1^2}{R_o^2/N} = \frac{NO_1^2}{R_o^2} \tag{12.59}$$

which is the same as Eq. 12.53, as desired. The modulation for a single interference component is obtained from (12.58) as

$$M_{inc} = \frac{2O_1 R_o N^{-1/2}}{NO_1^2 + R_o^2} . \qquad (12.60)$$

Again letting $A_t^2 = NO_1^2$, we obtain

$$M_{inc} = \frac{2A_t R_o}{N(A_t^2 + R_o^2)} . \qquad (12.61)$$

A comparison of Eqs. 12.56 and 12.61 shows that for coherent superposition, the modulation varies as $N^{-1/2}$ while for incoherent superposition it varies as N^{-1}. Except for nonlinearities of the recording medium, the amount of light flux diffracted from the recording medium, the amount of light flux diffracted from the hologram is proportional to the square of the modulation. Thus for coherent superposition the signal flux is proportional to N^{-1}.

On the other hand, for incoherent superposition the signal flux varies as N^{-2}, which means that the number of bits that can be recorded in this way is severely limited.

Note that for coherent superposition, each of the N holograms contribute a proportion N^{-1} to the total diffracted flux so that the total diffracted light is independent of N. For incoherent superposition, the total diffracted flux goes as N^{-1}.

The foregoing clearly indicates that the best possible way to store information in a hologram is coherently, that is, all the information is stored simultaneously. This is not always possible, of course, since the input (object) plane may not be large enough to fit all of the information simultaneously. Whenever possible, though, it is desirable to record the information in a single exposure.

12.7.3   Storage Capacity

To obtain an order-of-magnitude estimate of the storage capacity of a hologram, we proceed as follows: Let a diffuse object D be placed a distance f from a lens, with the hologram plane a distance f behind the lens. We allow a small hole in the center of the object for the reference beam source (Fig. 12.29). We consider the object to consist of a large number of just resolvable points. We now inquire, What is the greatest number N_m of points that can be recorded on a recording medium which can only record spatial frequencies up to v_c c/mm?

To store the maximum number of points, the object D must be made as large as possible. The relationship between object size and recorded spatial

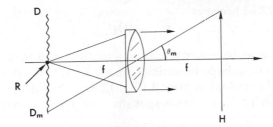

Fig. 12.29   A hologram recording arrangement useful for calculating the information storage capacity of a hologram.

frequency is

$$\nu = \frac{\sin \theta}{\lambda} = \frac{D}{f\lambda} \tag{12.62}$$

so that the maximum allowable object dimension D_m is given by

$$\nu_c = \frac{\sin \theta_m}{\lambda} = \frac{D_m}{f\lambda} \tag{12.63}$$

The hologram must be at least as large as D_m, and the lens is assumed to be larger than this so that we will not be concerned with the problem of vignetting.

The information stored in the hologram is read out (Fig. 12.30) by illuminating it with a plane wave and forming the image with a lens of focal length f. If the holographic system is diffraction limited, the spot size of the image of a single object point given by

$$ss = \frac{f\lambda}{H}. \tag{12.64}$$

Therefore the maximum number of resolvable points in the image is

$$N_m = \frac{\text{Area of image}}{\text{Area of point image}} = \frac{D_m^2 H^2}{f^2 \lambda^2}, \tag{12.65}$$

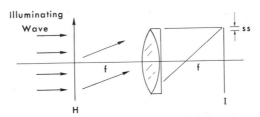

Fig. 12.30   Demonstrating the system for readout of the information stored in the hologram recorded with the system of Fig. 12.29.

which by Eq. 12.63 becomes

$$N_m = v_c^2 H^2. \tag{12.66}$$

Assuming that we have used a recording medium with the largest available resolving power (v_c as large as possible), we will want to use this same material for recording the readout in plane I (Fig. 12.30). Thus the material must be capable of resolving the minimum spot size ss $= f\lambda/H$, so that we must have

$$\frac{H}{f\lambda} \le v_c = \frac{D_m}{f\lambda}. \tag{12.67}$$

Hence we should have $H = D_m$. If $H > D_m$, then we would be able to store more information but would lose on the readout since all of the signal points would not be resolved. If we make $H < D_m$, we will unnecessarily restrict N_m since the minimum resolvable spot size will be larger than necessary.

The number N_m given by Eq. 12.66 is the maximum number of points (bits of information) we can store on a single two-dimensional hologram of dimension H. We must now determine the signal-to-noise ratio S/N. All recording media containing a stored signal will scatter some of the incident light during readout. It is this scattered light that we shall call the noise of the process.

In holography, readout is often accomplished by illuminating the hologram with laser light, or at least very monochromatic light, so that the scattered light, or noise, is coherent with the signal. This fact complicates somewhat the problem of defining a meaningful signal-to-noise ratio. Following the treatment of Chapter 8, we take the signal-to-noise ratio as the deterministic image irradiance I_i to the standard deviation ($\sigma^2 = \overline{[I - \bar{I}]^2}$) of the total irradiance:

$$\frac{I_i}{\sigma} = \frac{I_i}{I_N} \left[1 + \left(\frac{2I_i}{\bar{I}_N} \right) \right]^{-1/2} \tag{8.2}$$

where $\bar{I}_n$ is the average irradiance of the scattered light (noise) in the image plane.

We first determine the noise, $\bar{I}_n$. This is easily done with the aid of the curves in Figs. 8.2–8.5. These data give the light flux F_N scattered into a two-dimensional bandwidth B, per unit of incident flux F_o:

$$S = \frac{F_N/B}{F_o} \tag{12.68}$$

From the geometry of Fig. 12.30, we see that because of diffraction there is a range of angles $\Delta\theta$ of diffracted light that contributes to the image spot. Thus

$$ss = f\,\Delta\theta. \tag{12.69}$$

The range of spatial frequencies corresponding to $\Delta\theta$ is simply given by (see Eq. 12.62)

$$\Delta\nu = \frac{\cos\theta}{\lambda}\Delta\theta \approx \frac{\Delta\theta}{\lambda} \qquad (12.70)$$

so that by combining Eqs. 12.69 and 12.70 we have

$$B = \pi\,\Delta\nu^2 = \frac{\pi(\mathrm{ss})^2}{f^2\lambda^2}. \qquad (12.71)$$

This is the two-dimensional bandwidth that determines the amount of light scattered into the image. Hence

$$\frac{F_N}{B} = \frac{F_N\lambda^2 f^2}{\pi(\mathrm{ss})^2} = \bar{I}_N\lambda^2 f^2 \qquad (12.72)$$

since the irradiance $\bar{I}_n$ is just the flux per unit area. But by Eq. 12.68, we find that

$$\bar{I}_N = \frac{F_o S}{\lambda^2 f^2} \qquad (12.73)$$

where S is the scattering ratio plotted in Figs. 8.2–8.5.

The image irradiance I_i can be determined with a knowledge of the diffraction efficiency η. Thus

$$I_i = \frac{F_o\eta}{N_m\pi(\mathrm{ss})^2}. \qquad (12.74)$$

Knowing both I_i and $\bar{I}_N$ we can compute a signal-to-noise ratio using Eq. 8.2.

A reasonable estimate for the upper bound of the information capacity N_m/H^2 and the signal-to-noise ratio can be determined as follows: we suppose that $H = 10$ cm and $\nu_c = 10^4$ cycles/cm (1000 c/mm). Equation 12.66 then gives $N_m = 10^{10}$ as the maximum number of point object holograms that can be stored, provided that the system is diffraction limited and no space is allowed between images. The information capacity is therefore 10^8 bits/cm², which is the same theoretical capacity as for conventional techniques.

To determine the signal-to-noise ratio of this system, we recast Eq. 12.74 as

$$I_i = \frac{F_o\eta}{N_m\pi(\mathrm{ss})^2} = \frac{F_o\eta H^2}{N_m\pi f^2\lambda^2} = \frac{F_o\eta H^2 f^2\lambda^2}{D_m^2 H^2\pi f^2\lambda^2} = \frac{F_o\eta}{\pi D_m^2} = \frac{F_o\eta}{\pi H^2} \qquad (12.75)$$

for the optimum case of $H = D_m$. Hence

$$\frac{I_i}{\bar{I}_N} = \frac{F_o\eta}{\pi H^2}\frac{\lambda^2 f^2}{F_o S} = \frac{\lambda^2 f^2\eta}{\pi S H^2} = \frac{\eta}{\pi S\nu_c^2}. \qquad (12.76)$$

Taking $\nu_c = 10^3$ c/mm, $\eta = 0.02$, and $S = 1.0 \times 10^{-9}$ (c/mm)$^{-2}$ (cf. Fig. 8.2,

1000 c/mm and $T_a \cong 0.60$), we find

$$\frac{I_i}{I_N} = 6.37 \qquad (12.77)$$

and so

$$\frac{I_i}{\sigma} = 1.72. \qquad (12.78)$$

This number is probably too low for practical purposes but it does give an indication that signal-to-noise ratios exceeding unity are possible in holographic storage systems operating at a theoretical maximum capacity determined by diffraction. Of course, higher values of η can be achieved by bleaching, and, furthermore, the number of recorded images can be reduced, both of which will increase the signal-to-noise ratio.

Thus far we have not surpassed the storage capabilities of photographic film used in the conventional manner; however, there is still a third dimension to be considered. Recall that in Chapter 4 we discussed the effects of emulsion thickness on the properties of the hologram. There we found that diffraction is governed by the Bragg condition: both the grating equation and the law of reflection must hold simultaneously with regard to the directions of the illuminating and diffracted waves. In particular, it was found that for thick recording media, only a small misorientation of the illuminating beam is required to extinguish the image. For recording media of the order of 1 mm thick, such as the photochromics, this angular orientation tolerance may be as small as a few seconds of arc. Just how many holograms can be stored on a single plate in this manner is difficult to determine. The signal-to-noise ratio 12.78 will certainly fall off with each additional hologram, but just how fast the S/N declines has not been determined. In a noiseless system, with a cutoff frequency of 10^3 c/mm and an orientation tolerance of 10 sec of arc, the ultimate storage capacity on a 10×10-cm plate might be 3.6×10^{14} bits of information:

$$\frac{\text{Angular range}}{\text{Orientation tolerance}} \times N_m = \frac{100°}{(1/360)°} \times 10^{10} = 3.6 \times 10^{14}.$$

Even if only 10 maximum capacity holograms could be stored on a single plate with a reasonable S/N level, the gain over existing information capacity limits is some $10\times$, a not insignificant factor.

12.8   ULTRASONIC HOLOGRAPHY

Ultrasonic, or acoustical, holography is a variant on the general method of holography. The recording and readout techniques so far applied have been rather complex and cumbersome and the results have been of very poor

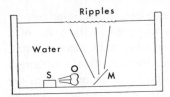

Fig. 12.31 Schematic representation of a possible ultrasonic hologram recording arrangement. The source of ultrasound S illuminates an object O and a mirror M directly. These two waves interfere at the water surface, forming the ripples. The ripple pattern is recorded in same manner, and this recording constitutes the hologram.

quality compared to what has been achieved with light. However, there is a lot of effort in this area and there is little doubt that simpler and more reliable recording techniques will evolve, with the subsequent improvement in results. The vast number of interesting applications for acoustical holography warrant the already sizeable effort in this area, and certainly some of these possible applications deserve mention in this section.

Most of the principles pertaining to light holography that have been discussed in the preceding chapters also apply to ultrasonic holography. Ultrasonic holography is a two-step process wherein the diffraction pattern of an object irradiated with ultrasonic waves is interfered with a mutually coherent reference wave. The resulting spatial irradiance distribution is recorded in some manner. Next the acoustical hologram is illuminated with a beam of light. The resultant diffraction from the hologram provides a generally reduced object wave, which can be used to form a three-dimensional visual image of the object.

The geometrical arrangements for making ultrasonic holograms are quite straightforward. One possible arrangement is shown in Fig. 12.31. The ultrasonic transducer S irradiates the object O and the resulting scattered wave is reflected to the water surface by means of a plane mirror. A portion of the acoustic wave is intercepted by the plane mirror M and directed towards the water surface as a reference wave. At the water surface, the acoustical object and reference waves interfere in a manner characteristic of the object O. This interference pattern appears as ripples on the surface of the water. The problem of suitably recording this pattern has been approached in several ways, but none to date have been very satisfactory.

One method that has been used requires photographing the ripple pattern on a water surface directly by means of a Schlieren system [39]. Other possible schemes include recording the waves on the water surface directly with a pressure-sensitive recording medium [40], scanning the acoustic field with an electrical transducer [41–43], or by driving the source hard enough to induce cavitation on a water surface and using the resulting localized turbulence to process photographic film [44]. With ultrasonics, it is not a necessary requirement for coherence that both object and reference beams originate from the same source. One can just as well use separate drivers for the object and reference beams, as long as they are mutually coherent.

The final image formation is achieved with more or less conventional means. The acoustical hologram, assuming it has been recorded in the form of a transparency, is illuminated with a collimated monochromatic beam of light. Since this wavelength is generally much smaller than the acoustical wavelength used to record the hologram, good imagery can be achieved only when the hologram is reduced in size by a fraction equal to the ratio of the optical to the acoustical wavelengths. The images are then formed in the usual manner.

The principal feature of ultrasonic holography is its ability to form three-dimensional images of objects that are opaque to light waves but transmit ultrasonic waves. Thus the interior of solid parts might be examined optically. Because of this feature, underwater and underground surveillance and exploration may be possible. Also, acoustical holography may find important applications in the medical field. Parts of the human body, for example, that are transparent to the sound waves, may be rendered visible by means of ultrasonic holography, with the image presented in a much more useful and easily interpreted manner than the simple shadowgrams now produced with x-rays.

REFERENCES

[1] L. H. Tanner, *J. Sci. Instr.*, **43**, 81 (1966).

[2] K. A. Haines and B. P. Hildebrand, *Appl. Opt.*, **5**, 595 (1966).

[3] L. O. Heflinger, R. F. Wuerker, and R. E. Brooks, *J. Appl. Phys.*, **37**, 642 (1966).

[4] R. E. Brooks, L. O. Heflinger, and R. F. Wuerker, *IEEE J. Quant. Elec.*, **QE-2**, 275 (1966).

[5] R. E. Brooks, L. O. Heflinger, and R. F. Wuerker, *Appl. Phys. Letters*, **7**, 248 (1965).

[6] R. E. Brooks, R. F. Wuerker, L. O. Heflinger, and C. Knox, Paper presented at the International Colloquium on Gas dynamics of Explosions, Brussels, Belgium, Sept. 20, 1967.

[7] J. M. Burch, A. E. Ennos, and R. J. Wilton, *Nature*, **209**, 1015 (1966).

[8] E. B. Aleksandrov, and A. M. Bonch-Bruevich, *Sov. Phys.—Tech. Phys.*, **12**, 258 (1967).

[9] J. E. Sollid, *Appl. Opt.*, **8**, 1587 (1969).

[10] N. E. Molin and K. A. Stetson, *Optik*, **33**, 399 (1971).

[11] C. Froehly, J. Monnerer, J. Pasteur, and J. C. Vienot, *Optica Acta*, **16**, 343 (1969).

[12] A. E. Ennos, *J. Sci. Instrum.*, **21**, 731 (1968).

[13] T. Tsuruta, N. Shiotake, and Y. Itoh, *Optica Acta*, **16**, 723 (1969).

[14] N. L. Hecht, J. E. Minardi, D. Lewis, and R. L. Fusek, *Appl. Opt.*, **12**, 2665 (1973).

[15] N. Abramson, *Appl. Opt.*, **8**, 1235 (1969).

[16] N. Abramson, *Appl. Opt.*, **9**, 97 (1970).

[17] N. Abramson, *Appl. Opt.*, **9**, 2311 (1970).

[18] N. Abramson, *Appl. Opt.*, **10**, 2155 (1971).

[19] N. Abramson, *Appl. Opt.*, **11**, 1143 (1972).

[20] R. L. Powell and K. A. Stetson, *J. Opt. Soc. Am.*, **55**, 1593 (1965).

[21] K. A. Haines and B. P. Hildebrand, *Phys. Letters*, **19,** 10 (1965).

[22] J. Upatnieks, A. Vander Lugt, and E. N. Leith, *Appl. Opt.*, **5,** 589 (1966).

[23] A. Vander Lugt, *IEEE Trans. Inf. Theory*, **IT-10,** 139 (1964).

[24] C. B. Burckhardt, *Appl. Opt.* **6,** 1359 (1967).

[25] H. M. A. El Sum, *Reconstructed Wavefront Microscopy*, Ph.D. Thesis, Stanford University, November 1952.

[26] A. V. Baez, *J. Opt. Soc. Am.*, **42,** 756 (1952).

[27] E. N. Leith and J. Upatnieks, *J. Opt. Soc. Am.*, **55,** 569 (1965).

[28] R. F. van Ligten and H. Osterberg, *Nature*, **211,** 282 (1966).

[29] G. W. Ellis, *Science*, **154,** 1195 (1966).

[30] W. H. Carter and A. A. Dongal, *IEEE J. Quant. Elec.* **QE-2,** 44 (1966).

[31] D. Gabor, *Proc. Roy. Soc. (London), Ser. A*, **197,** 454 (1949).

[32] B. J. Thompson, J. H. Ward, and R. Zinky, *Appl. Opt.*, **6,** 519 (1967).

[33] C. Knox, *Science*, **153,** 989 (1966).

[34] R. F. van Ligten, *Proc. Soc. Phot. Inst. Eng.*, **15,** 75 (1968).

[35] W. H. Carter, P. D. Engeling, and A. Dougal, *IEEE J. Quant. Elec.* **QE-2,** 44 (1966).

[36] W. L. Anderson, *Proc. Soc. Phot. Inst. Eng.*, **15,** 159 (1968).

[37] E. N. Leith and J. Upatnieks, *J. Opt. Soc. Am.*, **56,** 523 (1966).

[38] J. W. Goodman, *Appl. Phys. Letters*, **8,** 311 (1966).

[39] R. V. Mueller and N. K. Sheridon, *Appl. Phys. Letters*, **11,** 294 (1967).

[40] J. D. Young and J. E. Wolfe, *Appl. Phys. Letters*, **11,** 294 (1967).

[41] K. Preston, Jr. and J. L. Krenzer, *Appl. Phys. Letters*, **10,** 150 (1967).

[42] A. F. Metherall, H. M. A. El Sum, J. J. Dreher, and L. Larmore, *Appl. Phys. Letters*, **10,** 277 (1967).

[43] G. A. Massey, *Proc. IEEE (Letters)*, **55,** 1115 (1967).

[44] P. Greguss, *J. Phot. Sci.*, **14,** 329 (1966).

Appendix Fourier Transforms with Lenses

A.0 INTRODUCTION

Several authors [1–3] have shown that within certain limitations (not always explicitly stated) a lens forms, in its back focal plane, the Fourier transform of the field distribution in the front focal plane. Since this property of a lens is used so often throughout this book, we present here our own derivation, so that symbols and phase conventions will be well defined. We will treat only the one-dimensional case.

A.1 ANALYSIS

The starting point in any diffraction problem is Huygens' principle, which, in one dimension (cylindrical waves) may be written as follows:

$$g(x_L) = \left(\frac{-i}{4\lambda}\right)^{1/2} \int_{-\infty}^{\infty} F(x_o)[1 + \cos \theta] \frac{e^{ikr}}{(r)^{1/2}} \, dx_o. \tag{A.1}$$

With this formulation, we assume that each elementary strip of a wavefront $F(x_o)$ acts as a line source of cylindrical waves that lead the parent wave in phase by $\pi/4$. The amplitude contributed by this element at some distant point at the proper later time is the wave amplitude of the parent wave times the factor

$$\frac{1 + \cos \theta}{2(\lambda r)^{1/2}}. \tag{A.2}$$

The factor $\frac{1}{2}(1 + \cos \theta)$ is the obliquity factor, with θ the angle between the line from dx_o to x_L and the normal to dx_o (see Fig. A.1). The length r is the

268

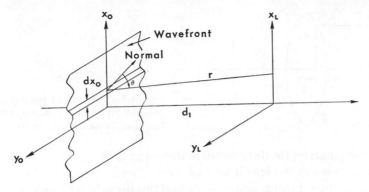

Fig. A.1   Notation and meaning of Eq. A.1.

distance between the strip dx_o and the point x_L on the lens at which we wish to know the field.

If we restrict the analysis to small fields, say $\theta \leq 10°$, then there is only a negligible error in writing

$$\tfrac{1}{2}(1 + \cos \theta) \approx 1 \tag{A.3}$$

and under this assumption we may also bring the factor $r^{-1/2}$ outside of the integral (A.1), as this will not change significantly as dx_o moves over the wavefront. Actually, there is a much more stringent requirement on θ imposed by the approximations that follow, so the approximation A.3 will always be quite good.

With these approximations, the statement of Huygens' principle becomes

$$g(x_L) = \left(\frac{-i}{\lambda d_1}\right)^{1/2} \int_{-\infty}^{\infty} F(x_o)e^{ikr} \, dx_o \tag{A.4}$$

where we have used $r^{-1/2} \cong d_1^{-1/2}$. The factor e^{ikr} describes the phase of the secondary wavelet. We now apply this formulation to the situation shown in Fig. A.2. The x_o, y_o plane will be taken as the object plane, and we suppose

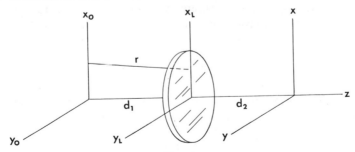

Fig. A.2   Notation and geometry for determining the effect of a lens on a wavefront.

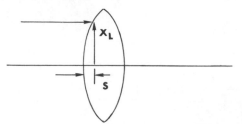

Fig. A.3 The optical path difference between an axial ray and a ray at a height x_o for an equiconvex lens.

that we can describe the disturbance in this plane as $F(x_o)$. The x_L, y_L plane is the plane in which the lens is located. Our ultimate goal is to find the disturbance in the x-y plane, and to show that this disturbance looks very much like the Fourier transform of $F(x_o)$.

We first write

$$r = [(x_o - x_L)^2 + d_1^2]^{1/2} \cong \pm d_1 \pm \frac{(x_o - x_L)^2}{2d_1}. \tag{A.5}$$

Since d_1 is a positive number, r will always be positive if we choose the upper signs of Eq. A.5. Thus the phase convention is that phase is advancing for the wave traveling left to right (positive z direction), and for $r \geq 0$, we simply write this as e^{ikr}. It is the approximation (A.5) that restricts the value of θ to only a few degrees (see [2] for a more detailed discussion of this approximation). We thus have for the disturbance just in front of the lens

$$g(x_L) = \left(\frac{-i}{\lambda d_1}\right)^{1/2} \int_{-\infty}^{\infty} F(x_o) e^{ikd_1} e^{ik(x_o - x_L)^2/2d_1} \, dx_o. \tag{A.6}$$

We know that the action of the lens must be such as to render a wave from a point source on the axis at $x_o = 0$ plane when $d_1 = f$, where f is the focal length of the lens. To do this, the lens must retard the phase of the wave more at the center than at the edges. The phase retardation factor can easily be derived for the equiconvex lens of Fig. A.3. When a plane wave is incident from the left, the optical path difference between the axial ray and a ray incident at a height x is just $2S(n - 1)$ for a lens with index n. The sag S is given by

$$S \cong \frac{x_L^2}{2R} \tag{A.7}$$

with the usual assumption that $R \gg S$, where R is the radius of curvature of the (equal) lens surfaces. The optical path difference is

$$\frac{(n - 1)x_L^2}{R}. \tag{A.8}$$

For a simple lens, the relation between the focal length and radius is

$$f = \frac{R}{2(n-1)} \tag{A.9}$$

so that the optical path difference may be written in terms of the focal length as

$$\frac{x_L^2}{2f}. \tag{A.10}$$

Thus the phase factor introduced by the lens of Fig. A.2 is just

$$e^{-ik(x_L^2/2f)} \tag{A.11}$$

and the disturbance (A.6), after passing through the lens, becomes

$$\psi(x_L) = g(x_L)e^{-ik(x_L^2/2f)}.$$

By repeating the application of Huygens' principle in the form (A.6), we can now calculate the disturbance in the x, y plane of Fig. A.2:

$$O(x) = \left(\frac{-i}{\lambda d_2}\right)^{1/2} \int_{-\infty}^{\infty} \psi(x_L)e^{ikd_2}e^{ik(x_L-x)^2/2d_2}\,dx_L$$

$$= \frac{-ie^{ik(d_1+d_2)}}{\lambda(d_1d_2)^{1/2}} \int\int_{-\infty}^{\infty} F(x_o)e^{ik(x_o-x_L)^2/2d_1}e^{ik(x_L-x)^2/2d_2}e^{-ik(x_L^2/2f)}\,dx_o\,dx_L$$

$$= \frac{-ie^{ik(d_1+d_2)}}{\lambda(d_1d_2)^{1/2}} e^{ik(x^2/2d_2)} \int\int_{-\infty}^{\infty} F(x_o)e^{ik(x_o^2/2d_1)}$$

$$\times \exp\left[i\frac{k}{2}x_L^2\left(\frac{1}{d_1}+\frac{1}{d_2}-\frac{1}{f}\right)\right] \exp\left[-ikx_L\left(\frac{x_o}{d_1}+\frac{x}{d_2}\right)\right]\,dx_o\,dx_L. \tag{A.12}$$

But

$$\int_{-\infty}^{\infty} \exp\left[i\frac{k}{2}x_L^2\left(\frac{1}{d_1}+\frac{1}{d_2}-\frac{1}{f}\right)\right] \exp\left[-ikx_L\left(\frac{x_o}{d_1}+\frac{x}{d_2}\right)\right]\,dx_L$$

$$= \left[\frac{\lambda}{-i(1/d_1+1/d_2-1/f)}\right]^{1/2} \exp\left[-i\frac{k}{2}\left(\frac{x_o}{d_1}+\frac{x}{d_2}\right)\Big/\left(\frac{1}{d_1}+\frac{1}{d_2}-\frac{1}{f}\right)\right], \tag{A.13}$$

so

$$O(x) = \left(\frac{-i}{\lambda}\right)^{1/2} e^{ik(d_1+d_2)} e^{ik(x^2/2d_2)}$$

$$\times \exp\left[-i\frac{k}{2}\frac{x^2(d_1/d_2)f}{f(d_1 + d_2) - d_1d_2}\right] \int_{-\infty}^{\infty} F(x_o)e^{ik(x_o{}^2/2d_1)}$$

$$\times \exp\left[-i\frac{k}{2}\frac{x_o^2(d_2/d_1)f}{f(d_1 + d_2) - d_1d_2}\right] \exp\left[-ik\frac{xx_o f}{f(d_1 + d_2) - d_1d_2}\right] dx_o.$$

$$(A.14)$$

We see that the quadratic exponentials under the integral will cancel if we choose d_1 and d_2 such that

$$\frac{(d_2/d_1)f}{f(d_1 + d_2) - d_1d_2} = \frac{1}{d_1}, \qquad (A.15)$$

which in turn implies

$$d_2 f = f(d_1 + d_2) - d_1d_2. \qquad (A.16)$$

Also, to make Eq. A.14 look like a Fourier transform, we must cancel the quadratic exponentials in x in front of the integral, or

$$d_1 f = f(d_1 + d_2) - d_1d_2. \qquad (A.17)$$

Equations A.16 and A.17 can hold simultaneously only for $d_1 = d_2$, and this implies $d_1 = d_2 = f$. Thus Eq. A.14 becomes

$$O(x) = \left(\frac{-i}{\lambda f}\right)^{1/2} e^{2ikf} \int_{-\infty}^{\infty} F(x_o)e^{-i(k/f)xx_o} dx_o. \qquad (A.18)$$

It is usual practice to neglect the constant phase factor e^{2ikf} since it is not of any importance. However, even without this factor, Eq. A.18 is still not in the form of a Fourier transform, so it is not necessary to neglect it. To obtain a true Fourier relationship, we first define a spatial frequency ν as

$$\nu \equiv \frac{x}{f\lambda}. \qquad (A.19)$$

Then Eq. A.18 becomes

$$O(\nu\lambda f) = \left(\frac{-i}{\lambda f}\right)^{1/2} e^{2ikf} \int_{-\infty}^{\infty} F(x_o)e^{-2\pi i\nu x_o} dx_o. \qquad (A.20)$$

By defining a new function

$$V(\nu) \equiv \left(\frac{-i}{\lambda f}\right)^{-1/2} e^{-2ikf} O(\nu\lambda f), \qquad (A.21)$$

Eq. A.20 becomes

$$V(v) = \int_{-\infty}^{\infty} F(x_o)e^{-2\pi i v x_o}\, dx_o, \qquad (A.22)$$

which is the desired Fourier transform relationship.

REFERENCES

[1] J. E. Rhodes, Jr., *Am. J. Phys.*, **21**, 337 (1953).

[2] E. N. Leith, A. Kozma, and C. J. Palermo, Notes from Int. to Opt. Data Processing, Eng. Summer Conferences, Univ. of Mich., Ann Arbor, Michigan, 1966.

[3] K. Preston, Jr., *Optical and Electro-Optical Information Processing* (Symposium Proceedings, Boston, 1964), J. T. Tippett, D. A. Berkowitz, L. C. Clapp, C. J. Koester, and A. Vanderburgh, Jr., Eds., MIT, Cambridge, 1965, p. 59.

Index

Aberrations, 5, 28, 39, 40, 169, 172, 173-177
  astigmatism, 173-177
  coma, 173-176
  distortion, 173-176
  field curvature, 173-176
  spherical, 173-176
Abramson, N., 232, 233
Achromatic fringes, 191
Amplitude transmittance, 7, 14, 15, 45, 94, 110, 111, 122, 133
Angular magnification, 172
Astigmatism, 173-177
Axial mode, 187

Baez, A. V., 6, 7, 248
Bandwidth, of source, 152-155, 248
  spatial frequency, 196-198
Beam Balance ratio, 81, 103, 111
Bessel function, 78, 79, 112
Bleaching, 77, 82, 87, 104
  reversal, 82, 86
Bloom, A. L., 178
Bragg, W. L., 3, 5
Bragg angle, 114, 122, 124, 202
  condition, 28-30, 57, 60, 62-64, 71, 92, 113-118, 120, 164
  planes, 96, 101, 102

Brooks, R. E., 187, 188, 190, 229
Buerger, M. J., 3
Burckhardt, C. B., 54, 77, 246

Carrier frequency, 6
Character recognition, 242-247
Chromaticity diagram, 105
Coburn, W., 210
Coherence, definition, 2
  length, 9, 21, 104, 187, 237
  of light, 8, 126
  requirement, 194
  of ruby laser, 190
  of source, 6, 237
  spatial, 185-187
  temporal, 178
Coherent superposition, 103
Collier, R. J., 92, 99, 213
Color center, 205
Color holography, 10, 27, 32, 75, 91-106
Colorimetry, 104-106
Coma, 173-176
Conjugate image, 21, 25, 27, 28, 41, 62, 75, 81, 107, 248
Conjugate wave, 14, 20, 40, 79
Contour generation, 236
Coupled wave, equations, 115, 117

275

theory, 113, 119
Crosstalk images, 91-101
Curran, R. K., 211

Delta function, Dirac, 74, 146, 158, 184
Denisyuk, Y. N., 10, 27, 30, 71, 73
Density, 123
Dephasing measure, 114, 115
Dichromated gelatin, 82-84, 93, 210, 213
Diffraction efficiency, 77, 80-83, 85, 87, 93, 98, 102, 104, 107-124, 132, 136, 202, 208, 210, 211, 213, 216, 239
Diffuse illumination, 21, 81, 133, 138, 139, 188
Direction cosines, 56, 60
Distortion, 173-176
Ditchburn, R. W., 194
Doppler linewidth, 179, 185
Double exposure interferometry, 227-229
Dyson, J., 5

Electro-optic crystals, 216, 217
Electron microscope, 4
Ellis, G. W., 251
El Sum, H. M. A., 6, 9, 248
Exposure, 45
  modulation, 57

Ferroelectric, 218
Field curvature, 173-176
Field of view, 158, 159, 163, 167, 197
Film grain noise, 125, 126-132
Fourier decomposition, 17
Fourier Transform, 3, 127, 140, 149, 150, 157, 162, 163, 165, 243, 268-273
Fourier Transform hologram, 26, 27,

43, 44, 46, 141, 144, 152, 154, 155, 156, 158, 163, 167, 169, 197, 198, 242
  lensless, 27, 47, 198
Fraunhofer approximation, 37
Fraunhofer hologram, 26, 29, 32, 37, 38, 42
Fresnel hologram, 25, 32, 34, 37, 38, 42, 44, 154, 159, 164, 198

Gabor, D., 3, 4, 5, 7, 9, 10, 14, 21, 247, 248
Gabor hologram, 9, 14, 21, 76, 188, 194, 196, 197, 250
Gamma, photographic, 7
Gas laser, 178-186
  spatial coherence, 185
  temporal coherence, 178
Gaussian image point, 95, 96, 171
Goodman, J. W., 130
Granularity, 125, 126
Greer, M. O., 132

Haine, M. E., 5
Haines, K., 210, 236
Hecht, N. L., 233
Heflinger, L. O., 190, 229
Hercher, M. M., 187
Hildebrand, B. P., 236
Holo-diagram, 232
Hologram, 1, 2, 13, 199, 200
  absorption, 76
  acoustical, 766
  amplitude, 109, 111, 117, 120, 123, 136
  bleached, 76, 77, 82, 83, 86-89
  color, 10, 27, 32, 75, 91-106
  computer, 112
  dichromated gelatin, 83, 210-213
  diffuse, 197
  Fourier Transform, 26, 27, 43, 44, 46, 141, 144, 152, 154-156, 158,

163, 167, 169, 197, 198, 242
Fraunhofer, 26, 29, 32, 37, 38, 42
Fresnel, 25, 32, 34, 37, 38, 42, 44,
   154, 159, 164, 198
Gabor, 9, 14, 21, 76, 188, 194, 196,
   197, 250
Leith-Uputniks, 36, 48
lensless Fourier Transform, 27, 47,
   198
Lippmann, 68, 71
off-axis, 8, 13, 19, 24, 27, 48, 194,
   198
   see also Leith-Uputniks
phase, 76-89, 112, 113, 115, 116,
   119, 133, 135, 201, 208, 213
<plane, 24, 27, 28, 33-47, 51, 97,
   109
reflection, 30, 68, 71, 101, 119-121
thick, 27, 28, 30, 36, 48-75, 85, 93,
   100, 104, 113, 122, 201, 202
thin, *see* Plane
volume, *see* Thick
white light, 30, 73, 75, 92, 98
   see also Lippmann
X-ray, 6
Holographic applications, 220-267
Holographic interferometry, 221, 238
   double exposure, 227
   Contour generation, 236
   multiple exposure, 234
Holography, acoustical, 139, 264
   basic description, 1, 13
   color, 41
   diffuse illumination, 8
   off-axis, 8, 13, 19, 24, 27, 48, 194,
      198
Huygens' principle, 268, 269, 271

Image, conjugate, 21, 25, 27, 28, 41,
   62, 75, 81, 107, 248
   crosstalk, 91-101
   primary, 21, 25, 27, 28, 63, 81,

107, 248
pseudoscopic, 25
real, 20, 25, 27, 28, 40
relief, 76, 82, 84, 87, 119
resolution, 152, 167, 238, 247
virtual, 20, 25, 27, 28, 40, 63
Index of refraction, 77, 82, 87, 114,
   117, 119, 202, 208
Information, capacity, 85, 260, 263
   storage, 27, 107, 206, 257
Interferometry, double exposure, 227-
   229
   holographic, 107, 221
   multiple exposure, 234
   single exposure, 222, 224, 225
   time average, 234

Jacobson, A. D., 187

Kirkpatrick, P., 6
Knox, C., 250
Kogelink, H., 113, 114, 115, 122
Kozma, A., 132
Kurtz, C. N., 88, 207

Lamberts, R. L., 88
Laser, 3
   gas, 8, 178
   Q-switched ruby, 190, 250
   ruby, 187, 188, 229
   solid state, 187
Lee, W. H., 132
Leith, E. N., 6, 7, 8, 10, 13, 16, 21,
   25, 54, 69, 91, 92, 95, 97, 194,
   247, 248, 250
Lin, L. II., 27, 75, 84, 92, 100
Line spread function, 145, 147, 155,
   157, 167
Lippmann, G., 10, 71
Liquid crystal, 218

Magnification, angular, 172

longitudinal, 172
transverse, 40, 170, 171, 248
Mandel, L., 92
Matched filtering, 242
McClung, F. T., 187
Meier, R. W., 25, 39, 84, 85, 215
Meyerhuffer, D., 211, 212
Microscopy, holographic, 247, 249, 254
Mode, 179-185
  axial, 187
  spatial, 186
  transverse, 186
Modulation transfer function, 109, 110, 155-158, 162, 163, 165, 167, 169
Multicolor wavefront, 96, 101
Multiple exposure interferometry, 231
Mulvey, T., 5

Noise, 80, 81, 83, 84, 87, 99, 107, 113, 121, 125-142, 216, 254
  nonlinearity, 135
  speckle, 136
Nonlinearity, 7, 19, 107, 112, 122, 126, 133, 135
  noise, 135

Object wave, 2, 13, 16, 21
Osterberg, H., 251

Pennington, K. S., 27, 75, 92, 99, 100
Phase, difference, 44, 50, 72, 181
  noise, 132
  variation, 37
  of wave, 4, 5, 8, 16, 19, 20, 39, 45, 75, 109, 138
Phase holograms, 76-89, 112, 113, 115, 116, 119, 133, 135, 201, 208, 213
Photoconductor, 84, 85, 126, 214, 215, 218

Photochromatic materials, 125, 132, 205, 206, 264
Photographic emulsions, 36, 48, 63, 76, 87, 119, 122, 130, 200, 202, 264
  bleached, 77, 82, 83, 93, 202
Photopolymers, 208, 209
Photoresists, 77, 207, 208
Plane hologram, 24, 27, 28, 33-47, 51, 97, 109
Point spread function, 147, 151
Powell, R. L., 234
Primary image, 21, 25, 27, 28, 63, 81, 107, 248
  wave, 39, 62, 79
Pseudoscopic, 25
Pupil function, 55, 56

Q parameter, 109
Q switched (laser), 90, 250

Ramberg, E. G., 54
Rayleigh, Lord, 138
Real image, 20, 25, 27, 28, 40
Recording, materials, 196-219
  effect on resolution, 196
  general requirements, 196
  for phase holograms, 82
Redundancy, 139-141
Reference wave, 2, 13
  coding, 98
  off axis, 7
Reflection hologram, 30, 68, 71, 101, 119-121
Relief image, 76, 82, 84, 87, 119
Resolution, 140, 141, 143, 151, 155, 163, 169, 196, 205, 208, 248
  image, 143-177
  of recording material, 196
Rogers, G. L., 5, 6, 9
Ruby laser, 187, 188, 229
Ruticon, 217

Scattered flux, 129, 130, 132
Scattering, 125, 126, 132
Sensitivity, 199, 202, 206
  spectral, 203, 204
Shankoff, T. A., 211
Sheridon, N. K., 217
Signal-to-noise-ratio, 81, 87, 98, 107,
    108, 130, 132, 257, 262-264
Single exposure interferometry, 222,
    224, 225
Spatial coherence, 185-187
  frequency, 17, 18, 44, 51, 52, 59,
    67, 82, 84, 87, 129, 152, 153,
    155, 181, 185, 196, 197, 198
  mode, 186
  multiplexing, 92, 97, 104
Speckle, 24, 25, 81, 82, 109, 126,
    133, 136, 138-141, 151, 224
Spread function, line, 145, 147, 155,
    157, 167
  line amplitude, 148, 150, 163
  point, 147, 151
Stetson, K. A., 234
Surface relief, 85, 201, 202, 208-210,
    215

Temporal coherence, 178
Thermoplastic, 36, 48, 76, 80, 82, 84,
    85, 93, 125, 126, 132, 213-216
Thompson, B. J., 248, 250
Time average interferometry, 234
Transmittance, amplitude, 7, 14, 15,
    45, 94, 110, 111, 122, 133
Twin wave, 4, 5, 7

Upatnieks, J., 6, 7, 8, 10, 13, 16, 21,
    91, 92, 95, 97, 140, 194, 239,
    247, 248, 250
Urbach, J. C., 84, 85, 215

Vander Lugt, A., 242
van Heerden, P. J., 10, 27
van Ligten, R. F., 251, 253
Vibration, 189, 234
Virtual image, 20, 25, 27, 28, 40, 63

Wavelength change, 62
  sensitivity, 68
Wiener, D., 187
Wiener spectrum, 127
Wuerker, R. F., 190, 229